Zoophysiology Volume 12

Coordinating Editor: D. S. Farner

Editors:

W. S. Hoar B. Hoelldobler

K. Johansen H. Langer G. Somero

E. Skadhauge

Osmoregulation in Birds

With 42 Figures

Springer-Verlag
Berlin Heidelberg New York 1981

Professor Dr. Erik Skadhauge
Institute of Medical Physiology, Dept. A
University of Copenhagen, The Panum Institute
Blegdamsvej 3c, DK-2200 Copenhagen N, Denmark

The front cover illustrates first the three Australian birds with which the author has worked – the emu, the galah, and the zebra finch; second, representatives of the two groups of marine birds used most often for osmoregulatory studies – ducks and gulls.

ISBN 3-540-10546-8 Springer-Verlag Berlin Heidelberg New York
ISBN 0-387-10546-8 Springer-Verlag New York Heidelberg Berlin

Library of Congress Cataloging in Publication Data. Skadhauge, Erik. Osmoregulation in birds. (Zoophysiology; v. 12) Bibliography: p. Includes index. 1. Osmoregulation. 2. Birds – Physiology. I. Title. [DNLM: 1. Birds – Physiology. 2. Water – Electrolyte balance. WI Z0615M v. 12/QL 698 S6260] QP90.6.S55   598.2'1   80-28792.

Typesetting, printing and bookbinding: Brühlsche Universitätsdruckerei, Giessen.
2131/3130-543210

Preface

The approach of this treatise is physiological throughout. In the eyes of the author it answers the rhetorical question raised by Maurice B. Visscher at the Physiology Congress in Washington D.C. in 1968: Does physiology exist? What he meant by this question was whether the fields of cellular physiology and physiology of the various organ systems had become so large that physiology as such had vanished. The firm answer is that physiology does indeed exist. Although it is important to study physiological problems at the subcellular level, it is important – and equally difficult – to study organ regulation at the cellular level, organ interaction, and integration into the whole organism. An account of avian osmoregulation from an integrated point of view is attempted in this book.

Since reading Homer W. Smith's *From Fish to Philosopher* and August Krogh's *Osmoregulation in Aquatic Animals* vertebrate osmoregulation has been in the center of the author's interest. The focus was set on avian osmoregulation after personal contact with the School of Krogh when working in the laboratory of Bodil M. Schmidt-Nielsen. The fundamental concepts and isotope techniques introduced by Hans H. Ussing have been of constant inspiration. An excellent example for the study of osmoregulation at the cellular level was given by the late Jean Maetz. The writing of this book was suggested by Donald S. Farner who is thanked for thorough editorial assistance, and especially with help in the subtle semantic peculiarities of the English language.

A large part of the author's experimental work has been carried out in laboratories abroad. Thanks are due particularly to Donald S. Bradshaw and Terence J. Dawson in Australia, for hospitality. The collaboration of Niels Bindslev, Bjarne G. Munck and David H. Thomas in the study of the lower intestine of the domestic fowl is gratefully acknowledged. Major grant support was received from the Danish Science Research Council and the NOVO Foundation.

V

Excellent secretarial assistance and drawing of figures was provided by Mrs. Dorthe Davidson, Bodil Søndergaard, Ulla Brat and Rita S. Christiansen.

Finally I wish to thank Poul W. Skadhauge for the cover picture, my family for forbearance, and Springer-Verlag for excellent cooperation.

Copenhagen, April 1981   ERIK SKADHAUGE

Contents

Chapter 1
Introduction 1

1.1 Problems of Osmoregulation 1
1.2 Review of Reviews 2
1.3 Anatomy of Body Fluids 3

Chapter 2
Intake of Water and Sodium Chloride 13

2.1 Thirst and Appetite for Salt 14
2.2 Problems of Interpretation of Preference
    and Discrimination Tests 15
2.3 Drinking of Fresh Water 17
    2.3.1 Total Water Deprivation 20
    2.3.2 Water Turnover in Carnivorous Birds 22
2.4 Salinity Tolerance. 23
    2.4.1 Intake of Saline Solutions by Birds
          with Salt Glands 26
2.5 Salt Requirements of Domestic Birds 26

Chapter 3
Uptake Through the Gut 30

3.1 Absorption of Salt and Water in the Small Intestine . 30
3.2 Osmotic and Solute Composition of Contents
    of the Small Intestine 31
3.3 Caecum: Role in Osmoregulation 32
    3.3.1 Presence of Caeca 32
    3.3.2 Flow of Material into the Caeca 34
    3.3.3 Mechanical Regulation of Inflow in the Caeca 36
    3.3.4 Quantitative Studies of Caecal Function . . . 36
    3.3.5 Faecal Water Content After Caecectomy . . . 38
    3.3.6 Role of Caecum in Avian Nutrition 38

Chapter 4
Evaporation 40

4.1 Evaporation from Resting Birds Under
    Thermoneutrality 40

4.1.1 Relationship Between Body Weight
and Oxygen Uptake 40
4.1.2 Relationship Between Body Weight
and Evaporation 42
4.1.3 Other Questions in Relation to Evaporation. . 46
4.2 Reaction to Heat Loads 49
4.3 Reaction to Dehydration 51

Chapter 5
Function of the Kidney 53

5.1 Excretion of Water and Salts 53
5.1.1 Renal Anatomy 53
5.1.2 Circulation of the Kidney 56
5.1.3 Renal Plasma Flow 58
5.1.4 Glomerular Filtration Rate (GFR) 59
5.1.5 Renal Excretion of Sodium and Water 62
5.1.6 Determinations of Single-Nephron Glomerular
Filtration Rates (SNGFR) 64
5.1.7 Renal Concentrating Ability 65
5.1.8 Renal Excretion of Chloride and Potassium Ions 71
5.1.9 Concentration of Several Ions in Ureteral Urine 71
5.1.10 The Basis of Hypertonicity
in the Medullary Cones 73
5.1.11 Correlation of Structure and Function 73
5.2 Action of Antidiuretic Hormones 75
5.2.1 Investigations of Mammalian Hormones . . . 76
5.2.2 Amount of Arginine Vasotocin
in the Neurohypophysis 77
5.2.3 Diabetes Insipidus 78
5.2.4 Renal Effects of Arginine Vasotocin 79
5.2.5 Metabolism of AVT, Plasma Concentrations,
and Effects on Plasma Parameters 82
5.3 Renal Nitrogen Excretion 84
5.3.1 Excretion of Uric Acid and Urates 85
5.3.2 Quantitative Role of Cation Trapping
in Precipitates of Uric Acid 90
5.3.3 Excretion of Urea 91

Chapter 6
Function of the Cloaca 92

6.1 General Aspects of Cloacal Function 92
6.1.1 Anatomy of the Cloaca 93
6.1.2 Storage of Urine in Coprodeum
and Large Intestine 97

6.1.3 "Anatomy" of the Droppings 98
6.1.4 Water Content of Faeces 99
6.2 Studies of Cloacal Absorption of Ureteral Urine . . 100
   6.2.1 Separation of Ureteral Urine and Cloacal Faeces 100
   6.2.2 Indirect Estimate of Cloacal Absorption of
       Ureteral Urine by Comparison of Normal Rate
       of Formation of Droppings and Flow Rate
       of Ureteral Urine 101
   6.2.3 Simultaneous Collection of Ureteral Urine
       and Faeces 102
6.3 In-vivo Instillation and Perfusion Studies 105
   6.3.1 Early Observations 105
   6.3.2 Basic Measurements of Transport Parameters
       of NaCl and Water of Coprodeum and Colon 107
   6.3.3 Changes of Parameters of Cloacal Transport
       Induced by Hydration and Dehydration and by
       Salt Loading and Depletion. 112
   6.3.4 Absorption/Secretion of K, NH_4, and PO_4 . . 114
6.4 In-vitro Studies of Transport Parameters
   of Coprodeum and Colon of the Domestic Fowl . . 115
   6.4.1 In-vitro Studies in the Galah 120
   6.4.2 In-vitro Studies in Ducks 121
   6.4.3 Comparison of In-vivo and In-vitro Experiments 121
6.5 Effects of Aldosterone on Cloacal Transport 121

Chapter 7
Function of the Salt Gland 125

7.1 Development of the Gland 125
7.2 Stimulus for Secretion 128
7.3 The Flow Rate and Ionic Composition
   of the Salt-Gland Fluid 135
7.4 Mechanism of Secretion 137
7.5 Adaptation and Hormones 139
7.6 Quantitative Role of Salt Gland in Excretion
   of an Acute Salt Load and Relation to Kidney/Cloaca
   in Salt/Water Balance 140

Chapter 8
Interaction Among the Excretory Organs 144

8.1 Birds Without Salt Glands 144
   8.1.1 Water and Salt Loading 144
   8.1.2 Dehydration 145
   8.1.3 Fractional Absorption/Secretion of K, NH_4
       and PO_4 149
8.2 Birds with Salt Glands 149

Chapter 9
A Brief Survey of Hormones and Osmoregulation 151

9.1 Corticosterone and Aldosterone. 151
9.2 Hypothalamo-Hypophyseal Hormones. 153
9.3 The Pituitary-Adrenal Axis; Renal
    and Extrarenal Secretions 154
9.4 A Note on the Renin-Angiotensin System of Birds . 154

Chapter 10
**Problems of Life in the Desert, of Migration,
and of Egg-Laying** 156

10.1 Desert Birds. 156
    10.1.1 The Zebra Finch. 157
    10.1.2 The Budgerigar 162
    10.1.3 Larks from the Namib Desert
           and Gambel's Quail 164
    10.1.4 The Ostrich 164
10.2 Problems of Osmoregulation During
     Migratory Flights 165
    10.2.1 Water Balance During Migratory Flights . . 165
    10.2.2 Calculations and Measurements of Energy
           Expenditure and Evaporative Water Loss
           During Migratory Flights 167
10.3 Osmotic Problems of Nestling Birds 169
    10.3.1 Water Transport by Sand Grouse Feathers . 170
10.4 Water and Salt Metabolism in Relation
     to Egg-Laying. 171
    10.4.1 Water Intake 171

References 175

Systematic and Species Index 199

Subject Index 201

X

Chapter 1

Introduction

1.1 Problems of Osmoregulation

The ions of Na and Cl are the major electrolytes of plasma and extracellular fluids, and constitute together with those of K and NH_4 the major fraction of osmolality of ureteral urine. The subject of osmoregulation, or osmotic and volume regulation, can be defined as the turnover and homeostasis of these ions and water.

The organism can be described as a flow system with three major components with respect to water and salts: first, intake of water and salts; second, uptake in the gut and production of metabolic water; and third, loss of water by evaporation from body surface and respiratory tract, and excretion of salts and water through the kidney, gut and, when present and functional, the salt gland. The sign of successful osmoregulation is constant osmolality of plasma and constant extracellular and intracellular volumes in spite of varying rates of intake, turnover, and losses.

In this book the quantitative relationships among the components of this flow system are described together with the mechanisms of uptake, retention, and excretion of salts, nitrogenous wastes and water by the organs involved in osmoregulation. After a description of the "anatomy" of body water and electrolytes follow chapters on intake of salt and water; uptake of these through the gut; evaporative loss of water; renal function; function of the cloaca; excretion by salt glands; interactions among organs; regulatory roles of hormones; and finally some special problems in survival in the desert, migration, and egg-laying. The approach is quantitative throughout and the book deals strictly with problems of avian physiology. Basic physiological principles, both regulatory and cellular, common to all or some classes of vertebrates, will only be treated in passing, and reference to other vertebrates will be given for comparison in a few cases.

Osmoregulation in birds poses a number of quantitative problems of functions of, and controlled interactions among organs. Drinking poses a problem: If fresh water is not available, how does salinity of available fluids relate to other parameters of osmoregulation? Evaporative cooling is a mechanism that reduces heat stress. What happens when a bird is both heat-stressed and dehydrated? Two problems involve the ability of the kidney to concentrate urine. The first is the formation of hyperosmotic urine during dehydration. As urine passes into the cloaca and is exposed to a resorptive epithelium, it would appear that the more concentrated the urine becomes during progressive dehydration, the greater the chance is that the osmotic work of the kidney is counteracted as more water would be dragged from the plasma into the cloacal lumen. The second problem is caused by the simultaneous presence of reptilian and mammalian nephrons, the former without, the latter with loops of Henle. The reptilian nephrons do not participate in the build-up of medullary hypertonicity and their glomeruli need not filter continuously. The mammalian nephrons should function at all times in order to permit delivery of salt to the counter-current system. How do these two nephron systems behave in the avian kidney under osmotic stress? There is also a problem that involves a major interaction among organs. Salt and water must be excreted together with uric acid in order to avoid clogging of the ureters with precipitated uric acid. This excretory system could continue to function, even during lack of salt and water, if salt and water were absorbed in the cloaca and salt excreted through the salt gland. This would save free water. Absorption of Na ions by the coprodeum is, however, reduced during salt loading.

Associated with the formation of concentrated urine is the problem of increase in concentration of solutes in ureteral urine as the renal absorption of water is increased. How does this process affect cloacal resorption? Finally, uricotelism is a necessary prerequisite for inflow of ureteral urine in the cloaca. Its absorptive epithelium is too permeable if the main end-product of nitrogen metabolism were urea. However, the concentrated precipitates of uric acid trap cations; but the fate of these ions in the cloaca is not yet known. How do birds handle lack or abundance of salt? These examples are sufficient to demonstrate the unresolved quantitative problems of interaction among organs in avian osmoregulation.

1.2 Review of Reviews

This section presents a compilation of surveys and reviews, entirely or partially concerned with osmoregulation in birds, published in the last two decades.

*Monographs.*All aspects of avian physiology have been dealt with by Sturkie (1976), endocrinology and osmoregulation by Bentley (1971), transport of renal and cloacal salt and water by Skadhauge (1973), and function of salt glands by Peaker and Linzell (1975).

*Textbooks on Avian Physiology.*Much information can be found in two recent advanced-level textbooks: *Physiology and Biochemistry of the Domestic Fowl* Vol. 1–2 (Bell and Freeman 1971) and *Avian Biology* Vol. 1–5 (Farner and King 1972–1975). The former contains chapters on intestinal absorption (Hudson et al. 1971), structure of the kidney (Stiller 1971), formation and composition of urine (Sykes 1971), the blood vascular system (Akester 1971), and composition of extracellular fluids (Freeman 1971). The latter has chapters on biology of desert birds (Serventy 1972), the blood-vascular system (Jones and Johansen 1972), respiration (Lasiewski 1972), digestion (Ziswiler and Farner 1972), osmoregulation and excretion (Shoemaker 1972), the endocrine glands (Assemacher 1973), and metabolic physiology of migration (Berthold 1975). In the *Handbook of Physiology* series thermoregulation of birds in humid heat was reviewed by King and Farner (1964), and birds in dry heat by Dawson and Schmidt-Nielsen (1964). Various aspects of avian physiology have been treated in the proceedings of a symposium on avian physiology edited by Peaker (1975).

Shorter Reviews. The adaption of birds to a hot, dry environment has been discussed in several reviews; first with emphasis on thermoregulation by Dawson and Bartholomew (1968) and by Dawson and Hudson (1970); second, with emphasis on water economy by Bartholomew and Cade (1963), and in granivorous birds by Bartholomew (1972); third, with emphasis on salt balance by Cade (1964); and further in relation to several aspects of desert life by Bartholomew (1964). The renal function of desert birds has been considered by Dantzler (1970), who has also reviewed the processes of excretion of uric acid (Dantzler 1978).

Control of excretion was surveyed by Peaker (1979), and the function of the adrenal cortex by Holmes and Phillips (1976). Neurohistology was reviewed by Oksche and Farner (1974).

1.3 Anatomy of Body Fluids

The following are of interest in relation to osmoregulation: (1) Total body water (TBW). (2) The rate of turnover of TBW. (3) The extracellular fluid volume (ECV) and the exchangeable sodium (Na_{ex}). (4) The plasma volume. (5) The plasma osmolality and electrolyte (NaCl) concentrations.

Estimates of Total Body Water (TBW). This parameter has been determined in a number of species (Table 1.1), either by direct weighing (before and after drying of the carcass to constant weight) or by antipyrine dilution. This drug is assumed to be freely distributed in the body water, thus marking not only extracellular, but also intracellular water. For the determination of TBW a water isotope is also usable. The three methods result in nearly the same values (see Tables 1.1 and 1.2). TBW is usually expressed as a fraction (per cent) of body weight. The problem of determination of body weight as such has been reviewed by Clark (1979). Regardless of the size of the bird the value determined on adult birds is around 60% of body weight. In several species the change of TBW from hatching to maturity has been measured. The results are identical: TBW was around 85% of body weight at hatching and decreases to about 60% as adult weight is attained. Both absolute values and change during maturation in birds are identical to the mammalian pattern. The young birds of some species grow to a weight greater than normal adult weight, and then decrease before fledging. Ricklefs (1968) has concluded, on the basis of measurements on nestling barn swallows (*Hirundo rustica*), that this weight recession is due entirely to decrease in water content.

Turnover of Total Body Water. The rate of turnover of TBW in birds as in other living beings can be determined with a water isotope, tritiated water being used most frequently. The isotope is injected into the animal, and after a brief equilibrium period a monoexponential fall of the tritium concentration in plasma or urine as function of time has been observed. From the monoexponential fall the

Table 1.1. Total body water

Species	Methods	Notes	% of BW	BW (g)	Author
Domestic fowl	Drying to constant weight	1 week	72	53	Medway and Kare (1957)
		4 weeks	68	224	
		32 weeks	57	1,771	
Domestic fowl	Antipyrine dilution	1 week	85	69	Medway (1958)
		4 weeks	69	221	
		32 weeks	55	1,619	
Domestic fowl	Antipyrine dilution	Hens	61	1,811	Weiss (1958)
Domestic pigeon	Drying to constant weight		64	362	LeFebvre (1964)
Tree sparrow (*Passer montanus*)	Drying to constant weight	2 days	84	5	Myrcha and Pinowski (1969)
		14 days	66	22	
Partridge (*Perdix perdix*)	Drying to constant weight	November	60	406	Szwykowska (1969)
		July	68	358	
Red-backed shrike (*Lanius collurio*)	Drying to constant weight	1 day	85	2.5	Diehl et al. (1972)
		15 days	69	28	
Meadow pipit (*Anthus pratensis*)	Drying to constant weight	0–4 days	83	< 5	Skar et al. (1972)
		3 weeks	67	18	
Starling (*Sturnus vulgaris*)	Drying to constant weight	1 day	85	12	Myrcha et al. (1973)
		3 weeks	69	70	

4

turnover rate can be calculated as the rate constant, or, more conveniently for comparison, expressed as the fractional exchange of body weight per unit time, usually per 24 h. If the injected amount of isotope has been measured, extrapolation of isotope concentration of body fluids to time zero (time of injection) permits calculation of TBW. The determination of turnover of tritiated water (HTO), and the estimates of total body water by this method, are reported in Table 1.2. An inverse relationship between turnover rate and body weight is apparent. This reflects the larger relative rate of metabolism, of drinking rate, and of evaporation in smaller birds (see Chaps. 2 and 4).

From a closer inspection of Table 1.2, three factors appear to be of interest in relation to water turnover: Physiological differences among species and within species; effect of egg laying; and of dehydration. First, the domestic duck has a very high turnover rate in relation to body weight. This may reflect the method of food ingestion of this species. Second, when comparison between the sexes has been made for the domestic fowl, hens have been observed to have a higher water turnover rate than roosters. As pointed out by Chapman and Mihai (1972), the water intake was increased beyond the amount of water contained in the eggs. Possible explanations for this relative polydipsia will be discussed in Sect. 10.3. Third, dehydration, as observed in the zebra finch (*Poephila guttata*) and in the emu (*Dromaius novaehollandiae*), reduced the rate of water turnover to less than half of that of normally hydrated birds. A substantial reduction was also observed in the roadrunner, (*Geococcyx californicus*), when it was only offered the water content of its prey.

The experimental determinations of TBW by isotope dilution (Table 1.2) are variable, but the values center around 62%–70% of body weight as observed by the methods described previously (see Table 1.1).

Too great an emphasis should probably not be placed on the very high or low values, as it is difficult to obtain a precise estimate of the counting efficiency for tritium, because of quenching, and the extrapolation to zero time also adds to the variation.

Extracellular Fluid Volume (ECV) and Exchangeable Amount of Sodium. (Na$_{ex}$). These parameters have been measured in selected species. Ruch and Hughes (1975) determined the ECV as the Br^{82} space. This ion is, like the Cl ion, largely distributed in the ECV. The authors investigated domestic ducks maintained on either fresh water (FW) or salt-water (SW), and both glaucous-winged gulls (*Larus glaucescens*) and roosters offered FW. The following values of ECV were found: Ducks (FW), 24.9% of body weight; ducks (SW), 26.4%, gulls, 38.2%; and roosters, 28.8%. The values for roosters and ducks are similar to those observed in mammals. The high value in glaucous-winged gulls is in agreement with the unusually high value of TBW obtained in this species (Table 1.2). In a study on white leghorn pullets Harris and Koike (1977) studied EVC as the iothalamate distribution volume, and the Na$_{ex}$ were determined when the birds received a purified diet with either a high or a low Na content. The low Na diet led to a decrease in ECV from 22.8% of body weight to 20.5%, and a decline of Na$_{ex}$ from 46.0 mequiv/kg body weight to 41.0 mequiv/kg, changes of 10% and 11% respectively. Significant changes in plasma Na or osmolality were not observed. Douglas (1968) observed an SCN (thio-

5

Table 1.2. Turnover rate and distribution volume (TBW) of tritiated water (HTO)

Species	Notes	HTO flux % of BW/day	TBW %	Body weight (BW) g	References
Zebra finch (*Poephila guttata*)	Hydrated	55	63	13	Skadhauge and Bradshaw (1974)
	Dehydrated	21	63	13	
Snowy plover (*Charadrius alexandrinus*)	Tap water, 25 °C	43	70	34	Purdue and Haines (1977)
	0.25 M NaCl	39	62	34	
Killdeer (*Charadrius vociferus*)	Tap water, 40 °C	46	67	71	
Semipalmated sandpiper (*Ereneutes pusillus*)	Tap water, 25 °C	55	59	24	
Japanese quail (*Coturnix coturnix*)	Male	22	67	105	Chapman and McFarland (1971)
	Female	20	62	117	
Burrowing owl (*Speotyto cunicularia*)	Single observation	7	42	140	
Orange-fronted parakeet (*Aratinga canicularis*)	Double observation	6	62	313	
Palm-nut vulture (*Gypohierax angulensis*)	Single observation	8	71	1,590	
Roadrunner (*Geococcyx californicus*)	Water ad lib. 8 °C	15	—	289	Ohmart et al. (1970)
	Fed mice, no water, 30–32 °C	9.0	64	292	
Glaucous-winged gull (*Larus glaucescens*)	Exposed to fresh water	6	81	763	Walter and Hughes (1978)
	Exposed to sea-water	6	77	817	
Domestic fowl	Roosters	6	64	2,600	Chapman and Black (1967)
	Hens	13	62	1,700	
Domestic fowl	Hens	5	57	2,150	Lopez et al. (1973)
	Pullets	9	76	268	
Domestic duck	Intact	23	62	3,060	Thomas and Phillips (1975b)
	Adrenalectomised	15	62	3,130	

Table 1.2. (continued)

Species	Notes	HTO flux % of BW/day	TBW %	Body weight (BW) g	References
Domestic duck	Fresh water	—	69	3,091	Ruch and Hughes (1975)
	Sea water	—	64	2,353	
Glaucus-winged gull (*Larus glaucescens*)	Fresh water	—	88	835	
Domestic fowl	Fresh water	—	54	2,506	Chapman and Mihai (1972)
Domestic fowl	Non-laying hens	7	54	3,490	
	Laying hens	12	62	4,900	
	Cocks	7	71	4,900	
Emu (*Dromaius novae hollandiae*)	Hydrated	6	63	32,700	Skadhauge et al. (1980c)
	Dehydrated	2	53	39,400	

Table 1.3. Plasma volume

Species	Method	Notes	Average value %(ml/100 g BW)	BW (g)	References
Domestic duck	RISA		6.9	300	Portman et al. (1952)
			6.5	900	
			5.5	1,550	
Diving ducks	T-1824		7.1	860	Bond and Gilbert (1958)
Dabbling ducks			6.4	980	
Domestic pigeon			4.4	310	
Domestic fowl		M	3.1	2,470	
Domestic fowl		F	4.4	1,900	
Red-tailed hawk (*Buteo jamaicensis*)		Immature	3.5	925	
Pintail (*Anas acuta*)	T-1824	F	5.4	620	Cohen (1967b)
Domestic fowl	T-1824	F	4.2	2,271	Wels et al. (1967)
Domestic goose	Cr⁵¹ + hematocrit	M	3.4	550	Hunsaker (1968)
		F	4.2	550	
Domestic duck	Inulin-C¹⁴, initial distribution volume	Fresh water	6.5	3,000	Bradley and Holmes (1971)
Domestic duck	T-1824		6.4	2,924	Stewart (1972)
Japanese quail	RISA	Young	5.9	75	Nirmalan and Robinson (1972)
		Adult M	4.7	98	
		Non-laying F	4.3	117	
		Laying F	4.7	138	
Domestic goose	T-1824	Control	6.0	4,240	Hanwell et al. (1971b)
		Salt loading	7.2	4,720	
Domestic duck	T-1824	Fresh water	5.8	3,091	Ruch and Hughes (1975)
		Sea-water	6.5	2,353	
Glacous-winged gull (*Larus glaucescens*)		Fresh water	7.2	835	
Domestic fowl			4.4	2,506	
Domestic fowl	T-1824	Control	4.8	800	Harris and Koike (1977)
		Low Na-diet	5.2	800	

8

cyanate) distribution space of 29.1% in the Adélie penguin (*Pygoscelis adeliae*), which is fairly high in comparison with SCN distribution space in mammals. Bradley and Holmes (1971) determined the inulin distribution space to be 16.7% in the domestic duck, a value that is typical for the space marked by inulin in mammals.

Plasma Volume. The circulating plasma volume is usually determined by a substance, such as T 1824 (Evans Blue), that binds to plasma proteins, or as the distribution volume of radioactively labelled serum albumin (RISA). Plasma volume may also be calculated indirectly from the red-cell volume, as determined by labelling with Cr^{51} and haematocrit. The plasma volume has been measured by these methods in several species of birds, both marine and terrestrial (Table 1.3). Birds of a fairly large range of body weight have been investigated, but information is lacking on species of less than 100 g. The determinations in marine birds range from around 5% to 7% of body weight whereas those of terrestrial species range from 3%–4% to 5%. A large number of species must be examined before firm conclusions concerning a possible difference can be drawn. A range around 5% is usually observed in mammals. A large blood volume might be an advantage for diving birds (Bond and Gilbert 1958). If the difference between marine and terrestrial birds is real, it may, however, be related to factors other than diving, as the highest value seems to be recorded in a gull (see Table 1.3). Conspicuous sex differences have not been encountered.

The erythrocyte volume is considered to be outside the scope of this review, but it may be noted that Newell and Shaffner (1950), who studied the domestic fowl from 400 g to 4000 g body weight found that the haematocrit is approximately 30% up to a body weight of 1800 g for both sexes. For higher body weights the haematocrit of hens remains around 30%, whereas that of roosters increases to nearly 50%.

Plasma Osmolality and Electrolyte Concentrations. These parameters have been reported in so many papers that a comprehensive compilation would be space consuming and redundant. A selective list of published values is presented in Table 1.4. The emphasis is placed on representation of a reasonable number of species, and illustration of the reaction to osmotic stress, particularly salt loading and dehydration. The plasma values of Table 1.4 are slightly higher (5%–10%) than those observed in most mammals. There appear to be no phylogenetic or ecologic relationship. Even considerable changes in the intake of water or NaCl can be handled by birds within a range of plasma osmolality and electrolyte concentrations of 1%–2%, a testimony to the efficiency of avian osmoregulation, which thus requires an error signal of only 1%–2% of concentration of the "milieu intérieur". As mentioned previously, the variation of ECV and Na_{ex} is around 10%. This regulatory capacity is also as effective as in mammals.

Osmoregulation develops only gradually after hatching (Grabowski 1967). The literature on the composition of eggs and the growing embryo has been summarised by Romanoff and Romanoff (1967).

Table 1.4. Plasma osmolality and electrolyte concentrations. (Fresh water ad lib. unless otherwise stated)

Species	Notes	Osmolality mOs	Na mequiv/l	Cl mequiv/l	References
Birds without salt glands					
Zebra finch (*Poephila guttata*)		336	149	123	Skadhauge and Bradshaw (1974)
	0.8 M NaCl	368	166	139	
Budgerigar (*Melopsittacus undulatus*)		336	147	106	Krag and Skadhauge (1972)
	Dehydration	349	156	108	
Galah (*Cacatua roseicapilla*)		336	142	117	Skadhauge (1974a)
	Dehydration	400	158	136	
Singing honeyeater (*Meliphaga virescens*)		343	158	127	
	Dehydration	384	174	140	
Crested pigeon (*Ocyphaps lophotes*)		336	149	123	
	Dehydration	370	169	140	
Emu (*Dromaius novae hollandiae*)		309	138	108	
	Dehydration	337	156	126	
Bobwhite (*Colinus virgianus*)		350	163	—	McNabb (1969b)
	Dehydration	395	162	—	
California quail (*Lophortyx californicus*)		336	157	—	
	Dehydration	374	151	—	
Gambel's quail (*Lophortyx gambelii*)		340	162	—	
	Dehydration	348	143	—	
Gambel's quail		378	178	—	Braun and Dantzler (1972)
California quail		355	—	—	Carey and Morton (1971)
	50% SW	477	—	—	
Gambel's quail	50% SW	349	—	—	
		456	—	—	
Savannah sparrows (*Passerculus sandwichensis beldingi*)	0.3 M NaCl	423	—	134	Poulson and Bartholomew (1962a)
	0.7 M NaCl	490–610	—	185–215	

Table 1.4 (continued)

Species	Notes	Osmolality mOs	Na mequiv/l	Cl mequiv/l	References
Savannah sparrow (*Passerculus sandwichensis brooksi*)	0.3 *M* NaCl	<400	—	<175	Poulson and Bartholomew (1972a)
Salt marsh sparrows (*Ammospiza spp.*)	0.2 *M* NaCl 0.5 *M* NaCl	346 383	— —	125 141	Poulson (1969)
Sage sparrow (*Amphispiza belli nevadensis*)	0.3 *M* NaCl	310 363	160 199	114 147	Moldenhauer and Wiens (1970)
Mourning dove (*Zenaidura macroura*)	Dehydration	372 410	176 190	136 142	Smyth and Bartholomew (1966a)
Domestic fowl	Dehydration Salt loading	312 341 338	151 163 166	119 130 142	Skadhauge and Schmidt-Nielsen (1967a)
Domestic fowl		319	146	120	Skadhauge (1967)
Domestic fowl		347	158	140	Skadhauge (1968b)
Birds with salt glands					
Domestic duck	2% NaCl	297 322	157	— —	Deutsch et al. (1979)
Domestic duck	Dehydrated	291 326	138 152	— —	Stewart (1972)
Domestic goose	NaCl load	297 322	142 154	— —	Zucker et al. (1977)
Black swan (*Cygnus atratus*)	0.36 *M* NaCl NaCl: 25 m*M*/kg	296 286 310	142 141 147	104 100 106	Hughes (1976)
Grey heron (*Ardea cinerea*)		317	201	145	Lange and Staaland (1966)
Jackass penguin (*Spheniscus demersus*)		298	153	106	Oelofsen (1973)

Table 1.4 (continued)

Species	Notes	Osmolality mOs	Na mequiv/l	Cl mequiv/l	References
Jackass penguin (*Spheniscus demersus*)		306	152	118	Erasmus (1978b)
American coot (*Fulica americana*)	NaCl: 34 mM/kg	259	135	97	Carpenter and Stafford (1970)
		348	189	133	
Guam rail (*Rallus owstoni*)	NaCl: 34 mM/kg	355	178	118	
		386	216	146	
Glaucous-winged gull (*Larus glaucescens*)	Sea-water	334	146	115	Hughes (1977)
		364	153	124	
Herring gull (*Larus argentatus*)	0.8 M NaCl	273	140	—	Ensor and Phillips (1972b)
		327	160	—	
Lesser black-backed gulls (*Larus fuscus*)	Dehydrated	321	139	—	
Roadrunner (*Geococcyx californicus*)		349	168	126	Ohmart (1972)

12

Chapter 2

Intake of Water and Sodium Chloride

The maintenance of the salt (mostly NaCl) content of the body requires regulation of intake, sensing of ionic concentrations, and regulation of excretion. Likewise, the maintenance of water content of the body requires regulation of intake, predominantly through thirst, sensing of concentration and possibly of total content, and regulation of loss. There is also evidence suggesting that there is some regulation of intake of NaCl. In this chapter principal emphasis is placed on quantitative studies concerning intake of water and NaCl. Studies on regulation by thirst, which seem to follow the same pattern as in mammals, and field observations on drinking behaviour, are referred to in passing.

When birds respond to osmotic challenges by the environment, when there is either a lack or surplus of NaCl or water, but also during temperature stress, in which there is a high rate of evaporation, there is an attempt to maintain levels of water and NaCl by matching intakes and losses. This means that in steady state the sum of water and NaCl intakes must equal the losses.

Water gains access to the organism via three routes: (a) drinking, (b) water in food, (c) water from metabolism. Water is lost by two routes: (d) evaporation, (e) reno-intestinal (cloacal) excretion. Marine and some other species lose, in addition, small amounts of water through the salt gland (f).

A bird is in water balance if the equation,

$$a + b + c = d + e + f,$$

is fulfilled. The relative importance of the components of this equation among species and in different physiological situations is essential to the study of osmoregulation. Just by inspection of the equation two important quantitative relations can be deduced.

1. For granivorous species, which obtain little preformed water in food, cloacal conservation of water is insufficient to maintain water balance when the rate of evaporation is high, and intake of free water becomes necessary. The rate of evaporation is high during heat stress because of evaporative cooling. The rate of evaporation is also relatively high in small birds because of relatively high rates of metabolism. Rates of drinking and evaporation will, therefore, be expected to increase in parallel for all birds during high ambient temperatures and, when these parameters are related to the body weight, with decreasing body weight. These expectations are sustained by results of experiments presented in this chapter and that on evaporation (see Sects. 4.1 and 4.2).

2. As already mentioned in Sect. 1.3, birds of prey receive a relatively large amount of preformed water in their food. They can, therefore, sustain a relatively

high rate of turnover without drinking free water (see p. 22). Even in the desert they are observed not to frequent water holes (Fisher et al. 1972).

For NaCl the requirement of steady state is fulfilled if the sum of reno-intestinal and salt-gland excretion equals the intake. However, with respect to intake of salt it is a major problem if no fresh water, but only solutions of higher salinity are available.

2.1 Thirst and Appetite for Salt

Thirst may be defined as the urge to drink. From human experience we know thirst as the perception of the need to regulate intake of water. There is no reason to doubt that voluntary drinking in other mammals and in birds is governed by a similar perception. As studied in recent years in mammals (see Andersson 1978 and Fitzsimons 1978), thirst can be induced by increase in concentration of NaCl in hypothalamic and juxtaventricular regions with angiotensin II as a contributing factor. In birds angiotensin II produces a dipsogenic response as studied in the domestic pigeon by Fitzsimons (1978), in the white-crowned sparrow (*Zonotrichia leucophrys gambelii*) (Wada et al. 1975), and in the Japanese quail (*Coturnix coturnix*) (Takei 1977). The preoptic area was found especially sensitive to mammalian angiotensin II. The Japanese quail, weighing 90–95 g, drank more than 10 ml water after injection of 1 µg into the preoptic area. "In these birds, the swelling of the crop sac was remarkable; and water flowed out of their mouths when they were caught by hand". (Takei 1977).

Kobayashi (1978) examined the sensitivity of 17 species of birds, and divided them into sensitive and insensitive groups. The former responded by drinking after an intraperitoneal injection of less than 500 µg angiotensin II per kg of body weight. The latter did not respond with drinking even after receiving 1 mg/kg body weight. The sensitive birds were neither desert occupants nor carnivorous, but the group of insensitive birds consisted of three birds of prey and the budgerigar (*Melopsittacus undulatus*). Kobayashi suggests that in these birds, which normally drink little water, the cerebral receptors to angiotensin II have decreased in number or affinity.

Concerning salt appetite in birds three types of evidence have been obtained: (1) Observations of salt eating in wild birds. (2) Laboratory experiments of salinity preference and discrimination. (3) Neurophysiological studies of salt-sensitive taste buds.

1. Salt eating has often been observed in granivorous birds (Van Tyne and Berger 1959, p. 258). Cade (1964) summarised the observations of salt eating in 14 species, mainly North American finches. The most often quoted source of salt was cattle salt blocks. It is reasonable to suggest that the finches eat salt for precisely the same reason for which NaCl is made available to the cattle: Both groups of animals receive a diet that is low in Na and high in K. The salt eating finches are most likely to suffer Na depletion.

2. Laboratory experiments to determine saline preference, tolerance, discrimination and threshold can be performed in three basic ways: (a) Single-

choice experiments may be conducted by offering one test solution at a time. This method is most suitable to determine salinity tolerance, as will be discussed in detail in the next section. (b) Two or more saline solutions may be present simultaneously. Only if osmotic and other problems (see below) do not complicate the issue, can this method determine preferred concentration, level of discrimination, and threshold. (c) To avoid osmotic problems when the birds have no access to free fresh water, the saline solutions may either be offered simultaneously with distilled water or the test solutions may be offered only for a small part of the day, with fresh water available for the remainder of the day.

3. The presence of saline-sensitive taste buds in domestic pigeons and fowl has been demonstrated by Kitchell et al. (1959); the NaCl-sensitive taste buds react to 0.2 M and higher concentrations of NaCl.

2.2 Problems of Interpretation of Preference and Discrimination Tests

Exposing birds without salt glands to NaCl solutions of increasing concentrations, as the only drinking fluid, has predictable consequences with respect to intake: First, as the osmolality of the fluid approaches the osmoregulatory capacity of the kidney (the renal concentrating ability), the bird drinks increasing volumes, so that by concentrating the fluid to the maximal osmolality of ureteral urine it can gain free water. Second, when the osmolality of the fluid approaches the maximal osmolality of ureteral urine the bird ceases, or reduces, consumption considerably. This is thus an indirect way of determining renal concentrating ability, or capacity for reno-intestinal conservation of water. It provides a rationale for the use of salinity tests to characterise xerophilic adaptation (see p. 24–26).

The experiments on the only species for which both drinking rates and osmolality of droppings and their electrolyte concentrations have been measured simultaneously for a wide range of salinities, the zebra finch (Skadhauge and Bradshaw 1974), amply confirm these considerations, but pose other problems. The fluid intake was augmented when saline of higher concentration was offered, until at 0.3 M NaCl the urine osmolality was close to the renal ceiling of approximately 1000 mOs (see Fig. 10.1). Calculations show that for 0.1–0.3 M NaCl the fluid intake and the osmolality of the droppings are such that the amount of free water being generated, approximately 4 ml per bird/day, is precisely identical to the normal intake of fresh water. The drinking rate thus seems to be determined totally by the osmotic need of the bird. From 0.4 M NaCl the intake declines exponentially to 0.8 M NaCl, which is the highest concentration that the birds will drink. The lesson from these findings is that single-choice or multiple-choice experiments, without access to fresh water, are likely to result in higher drinking rates with increasing salinity up to a maximum fixed by the concentrating capacity of the kidney, after which intake declines. Such findings can only be interpreted to indicate osmotic control of drinking. The NaCl-concentration at maximal drinking rate is not necessarily a "preferred" salinity. In several studies drinking rate does not increase over a wide range of saline concentrations. Even if the fluids had osmolalities below the osmotic ceiling of the kidney, such findings can neither be taken to indicate lack of preference nor absence of ability to discriminate, since the bird may be guided in its drinking pattern by osmotic clues. Furthermore, most such experiments have large standard errors reflecting large individual differences in responses. The danger of arriving at the false conclusion of "no difference" is high. To the already mentioned osmotic problems comes the consequence of augmenting the dietary load of NaCl. This may affect the renal cation trapping in urate precipitates (see p. 90) and affect the cloacal absorption rate (see p. 112). The increasing dehydration that occurs when higher salinities are offered will also affect these parameters.

Problems of interpretation exist also when one or more saline solutions have been offered together with distilled (or tap) water. First, when fresh water is offered in two-bottle tests against higher and higher concentrations of NaCl (or sea-water) in the range of 0.1–0.3 M NaCl both the sage sparrow (*Amphispiza belli nevadensis*) (Moldenhauer and Wiens 1970) and the house finch (*Carpodacus maxicanus*) (Bartholomew and Cade 1958) increase the relative fraction of freshwater intake. This is a rational osmotic response. When, however, Carey and Morton (1971) offered California and Gambel's quail (*Lophortyx californicus* and *Lophortyx gambelii*) tap water and 12.5%, 25%, and 37.5% of sea-water in four-bottle tests, they drank nearly equal amounts. Only for California quail was there significantly lower drinking of the 37.5% solution. Bartholomew and MacMillen (1961) offered California quail the choice between distilled water and 25% of sea-water of which the birds drank equal amounts.

The total intake of salt solution was, however, in the former study 100% (California quail) and 60% (Gambel's quail) higher than the intake of fresh water, whereas in the latter study the same total amount was consumed. There is thus a wide range of response patterns among various species and among different studies on the same species. Beyond the conclusion that birds in most cases behave rationally to osmotic clues, it is in the case of no difference in drinking impossible to judge whether the birds elect not to discriminate or whether they are insensitive to NaCl. Taken at face value, the discrimination level of these studies would seem to be around 0.2 M NaCl. When large differences in drinking are observed, it is impossible to judge whether it is an osmotic reaction or a taste response. The evidence is accordingly best presented as Bartholomew and MacMillen (1960) summarised their findings on the mourning dove (*Zenaidura macroura marginella*) with a purely observational description: "The birds did not discriminate between the different dilutions of sea-water on which they could maintain body weight, but they chose the least concentrated solutions available when kept on higher concentrations of sea-water".

To establish whether birds really display NaCl-taste responses, birds exposed to a very low salt diet should be offered the choice between distilled water and various low salinities. Of the rat it is known that only after NaCl depletion or adrenalectomy will the lowest threshold and level of discrimination be discovered.

They only study using a timed single-choice experimental system is that of Duncan (1962). He summarised earlier work on NaCl-taste in birds and studied the saline preference of the domestic pigeon with supplementary experiments on domestic fowl. The test solutions, including tap water as a reference, were offered for 2 h in the morning and tap water was always available for 4.5 h in the afternoon. No fluids were given for the rest of the 24 h. The pigeons and fowls showed a maximal intake of 0.085 to 0.1 M NaCl, and 0.14 M NaCl at the same level of intake as tap water. Higher concentrations were rejected. The discrimination level in this study was around 0.02 M NaCl in comparison to approximately 0.2 M NaCl in the two-bottle or multiple-choice experiments mentioned above (see p. 16), thus by a factor of 10 lower. Duncan's experiments presumably demonstrate the real ability of birds (domestic pigeon and fowl) to discriminate between saline solutions. As osmotic stress is absent, this ability must be interpreted as "taste", with a preference of about 0.1 M NaCl. Why the birds do not discriminate in the simultaneous choice experiments is not at all clear. Either they may not bother when water is available to avoid osmotic stress or they may actually prefer saline, and deliberately "dilute" with pure water. Only a detailed analysis of a large number of sip-pattern observations can clarify this.

Other problems emerge when drinking fluids are offered for only a short period. For example Harriman and Kare (1966a) performed a two-bottle 24 h drinking test (saline versus distilled water) on European starlings (*Sturnus vulgaris*), purple grackles *(Quiscalus quiscula)* and herring gulls *(Larus argentatus)*. At low concentrations the saline intake was not significantly different from 50% (random choice), but in all three species the NaCl consumption dropped considerably from 0.15 M and higher. The purple grackles were, however, still drinking 10% of the total intake from a 4 M NaCl solution. This implies that the average concentration of the fluid being drunk was 0.4 M NaCl. The total intake were not stated. For prolonged periods (Harriman and Kare 1966b) the purple grackles could not survice on 0.35 M NaCl. This makes the interpretation of the 24-h experiments difficult as they clearly reflect a non-steady behaviour. Since herring gulls, which live in brackish-water environment, showed freshwater preference, Harriman (1967) made a two-bottle test (NaCl versus distilled water) with marine laughing gulls *(Larus atricilla)*. Although this bird is able to survive on sea-water alone, because of the salt gland, it showed in this test no salinity discrimination for fluids less concentrated than 0.1 M NaCl (50% saline of total intake), but a gradual reduction of the saline intake from a 0.15 M solution (40% of total intake) to a 4 M solution (5% of total intake). This behaviour may be due to regulation of drinking by excretory capacity for salt (see p. 26).

In summary, the main regulator of drinking of saline solutions seems to be the osmotic requirement for free water. Refined experiments must be carried out to determine threshold, discriminating ability, and salinity preference.

2.3 Drinking of Fresh Water

In the Introduction the voluntary intake of free water was presented as a parameter that is governed to a large extent by the rate of evaporation which, in turn, is a function of body weight. The drinking rate has been measured in a number of

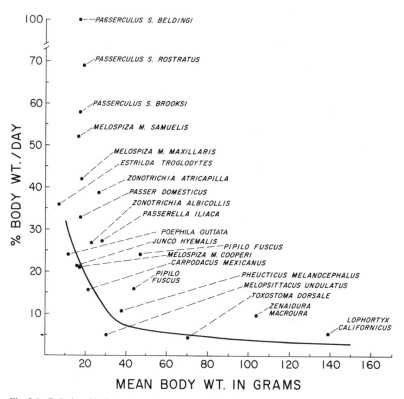

Fig. 2.1. Relationship between voluntary water intake and body weight. The relative intake of water at neutral ambient temperature is shown for a number of birds. The *solid line* indicates the evaporative water loss. Reproduced with permission from Bartholomew and Cade (1963)

largely seed-eating birds under ordinary laboratory conditions, i.e. ambient temperature of 20°–23°C and, usually uncontrolled, relative humidity of 50%–70%. The expectation of a parallel rise in water consumption, measured as a percentage of body weight, and evaporative water loss was clearly justified in a classical study of Bartholomew and Cade (1956). For body weights under 60 g water intake rose steeply with decreasing body weight. Subsequent work has fully confirmed this relationship. Data collected by Bartholomew and Cade (1963) are shown in Fig. 2.1. The general pattern is a water intake of about 5% of body weight per 24 h for birds weighing more than 100 g, and a water intake rising to a maximum, with certain exceptions, of about 50% of body weight per 24 h for birds weighing 10–20 g. As pointed out by Dawson and Bartholomew (1968), the relation between drinking rate (and evaporative water loss, see Chap. 4) and body weight is so important that it completely outweighs any other difference among species, most conspicuously those between desert and non-desert birds. It also surpasses the osmotically important parameters of renal concentrating ability. Exceptions to this general rule exist, however, in individual species (see below).

The measurements of voluntary drinking rates (Table 2.1) published in the decade following the survey of Bartholomew and Cade (1963; see Fig. 2.1) confirm

Table 2.1. Drinking rates. Birds on dry food and fresh water ad lib. at thermoneutrality

Species	Fresh water intake % of g BW/day	Minimal water intake % of g BW/day	Body weight (g)	References
Brewer's sparrow (*Spizella breweri*)	32	0	11	Ohmart and Smith (1970)
Tree sparrow (*Spizella arborea*)	30	15	15	
Zebra finch (*Poephila guttata*)	28	0	13	Skadhauge and Bradshaw (1974)
Stark's lark (*Spizocoryx starki*)	13	0	18	Willoughby (1968)
Grey-backed finch lark (*Eremopterix verticalis*)	8	0	18	
Sage sparrow (*Amphispiza belli nevadensis*)	49	12	19	Moldenhauer and Wiens (1970)
Vesper sparrow (*Pooecetes gramineus*)	20	—	28	Ohmart and Smith (1971)
Budgerigar (*Melopsittacus undulatus*)	7.8	0	29–46	Krag and Skadhauge (1972)
Cowbird (*Molothrus ater obscurus*)	22	3.4	28–38	Lustick (1970)
Redwinged blackbird (*Agelaius phoeniceus*)	20	10	35	Hesse and Lustick (1972)
Red crossbill (*Loxia curvirostra*)	16	—	30	Willoughby (1968)
Starling (*Sturnus vulgaris*)	54	0	68	Harriman and Kare (1966)
Purple grackle (*Quiscalus quiscula*)	54	0	88	
Herring gull (*Larus argentatus*)	20	0	814	
Japanese quail (*Coturnix coturnix*)	21	—	116	Chapman and McFarland (1971)
California quail (*Lophortyx californicus*)	6.1	1.5	149	McNabb (1969a)
Gambel's quail (*Lophortyx gambelii*)	7.5	1.3	149	
Bobwhite (*Colinus virgianus*)	7.7	3.4	181	
California quail	3.5	1.4	163	Carey and Morton (1971)
Gambel's quail	3.8	1.6	147	
Domestic duck	4.3	—	2,706	Fletcher and Holmes (1968)
Domestic fowl	3.0	—	3,150	Dicker and Haslam (1972)

the observed correlation between drinking rate and body weight. Table 2.1 presents mean values. The original publications usually show fairly large standard errors. This stems from the fact that among individual avian species, as also in mammals, there are habitual drinkers and non-drinkers.

For several species of birds both ad-libitum water intake and minimum water requirement of maintenance of body weight have been measured. The latter is usually one-third to two-thirds of the former. Some species can live without liquid water when fed dry seeds alone, as indicated by a zero for minimal water requirement in Table 2.1. This ability is associated with at least two adaptations, a relatively low rate of evaporation and a high renal concentrating ability as found in the budgerigar and the zebra finch. Water budgets of these and other non-drinking desert birds will be discussed in Sect. 10.1 and below (see p. 21). When birds of the same species, such as zebra finch, budgerigar, Gambel's quail, California quail, and mourning dove, have been studied by different investigators, results within the same order of magnitude have been found. The only results that deviate unusually from the general pattern are those of the study of Harriman and Kare (1966b), in which some unnoticed environmental factors may presumably explain the deviations. The extremely high intake of distilled water by the Savannah sparrows (*P.s. beldingi, rostratus*, and *brooksi*), studied by Cade and Bartholomew (1959), was suspected by these authors to be caused by lack of salt. Such moderate polydipsia—thirst induced by lack of salt—is known to occur in mammals. This explanation may find support in the observation of a 67% higher intake of distilled water than of tap water in one group of red-winged blackbirds (*Agelaius phoeniceus*) (Hesse and Lustick 1977).

In general, however, lack of correlation for seed-eating birds between ad-libitum water intake and any other parameter than body weight, including desert occupancy, is characteristic. As to be expected, xerophilic birds such as zebra finch, budgerigar, and the larks of the Namib desert (Willoughby 1968) have relatively low rates of voluntary water intake, but not outside the general range of their weight group. Voluntary drinking rate is thus a relatively weak parameter for characterisation of xerophilia and physiological adaptations thereto. This is largely due to two factors: First, xerophilia is much better measured as reaction to total water deprivation (see below) in which the well-adapted birds clearly stand out. Second, desert occupancy also depends on mobility, which gives access to distant sources of water. As pointed out by Bartholomew and MacMillen (1960), the mourning dove resists water deprivation relatively poorly, but it is a powerful flyer. It therefore occupies the same desert niche as, for example, Gambel's quail which is a poor flyer, but osmotically better adapted (Carey and Morton 1971) (see below).

2.3.1 Total Water Deprivation

The reaction to total water deprivation of granivorous birds fed dry seeds (water content around 10%) in laboratory experiments at normal temperature and relative humidity may be one of two. They either, usually after an initial weight loss, maintain body weight and live for months, or they continue to loose weight and die after a shorter or longer period during which the food intake declines. The body

Table 2.2. Birds that can survive on a diet of dry seeds without drinking. (From Bartholomew 1972)

Savannah sparrow *(Passerculus sandwichensis)*
Black-throated sparrow *(Amphispiza bilineata)*
Brewer's sparrow *(Spizella breweri)*
Vesper sparrow *(Pooecetes gramineus)*
Lark-like bunting *(Fringillaria impetuana)*
Zebra finch *(Poephila guttata)*
Red-headed finch *(Amadina erythrocephala)*
Cut-throat finch *(Amadina fasciata)*
Silver-bill *(Lonchura malabarica)*
Scaly-feathered weaver *(Sporopipes squamifrons)*
Stark's lark *(Spizocoryx starki)*
Grey-backed finch-lark *(Eremopteryx verticalis)*
Horned lark *(Eremophila alpestris)*
Budgerigar *(Melopsittacus undulatus)*

weight at death reflects the length of survival. It is relatively lower for the species that survive longer. The initial rate of weight loss also characterises the degree of xerophilia.

Bartholomew (1972) provided a list of 14 species (Table 2.2) that survive without drinking. Several of these are desert dwellers. In a number of these species the renal concentrating ability (see Chap. 5.5) and relative rate of evaporation (see Sect. 4.1) have been measured.

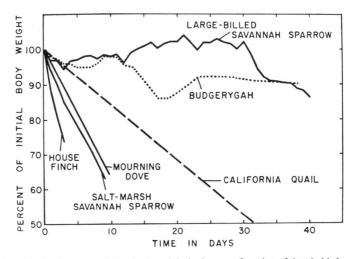

Fig. 2.2. Weight loss during dehydration. The relative body weight is shown as function of time in birds kept on a diet of dry seeds. The wide variation among species in the ability to withstand dehydration is apparent. The end-point for house finch *(Carpodacus mexicanus)*, salt-marsh savannah sparrow *(Passerculus sandwichensis beldingi)*, mourning dove *(Zenaidura macroura)*, and California quail *(Lophorticus californicus)* represent values at death. Reproduced with permission from Cade (1964)

21

Table 2.3. Weight loss during water deprivation. (From McNabb 1969a); Mean $\pm$ S.E. are reported

		Bobwhite (*Colinus virginianus*)	California quail (*Lophortyx californicus*)	Gambel's quail (*Lophortyx gambelii*)
Habitat		Humid	Mesic	Xeric
Initial rate of dehydration	(%/day)	4.6 ± 0.3	4.3 ± 0.5	3.8 ± 0.6
Mean rate of dehydration	(%/day)	3.7 ± 0.1	2.8 ± 0.3	2.2 ± 0.3

It will be seen that the ability to withstand dehydration is reflected, as expected, in the other physiological parameters that characterise ability to conserve water, the maximal osmotic urine to plasma ratio, and the rate of evaporative water loss as related to oxygen uptake.

Loss of body weight during water deprivation may best be displayed graphically (Fig 2.2), which shows the remarkable range in physiological responses to dehydration. Very similar data to those quoted by Cade for the California quail were obtained by McNabb (1969a) and by Carey and Morton (1971). McNabb compared the bobwhite (*Colinus virginianus*) to California (*Lophortyx californicus*) examples of closely related species which generally occupy humid, mesic, and xeric areas. For both mean and initial weight loss a gradient related to xerophilia is observed (Table 2.3). Water budgets of dehydrated desert birds will be further examined in Sect. 10.1.

2.3.2 Water Turnover in Carnivorous Birds

To survive birds of prey must obtain food that usually contains around 60% of preformed water. This makes them largely independent of free water although they drink occasionally when dehydrated. Bartholomew and Cade (1963) quoted observations on several owls, falcons, and hawks that could maintain weight in captivity without access to water when fed a diet of flesh. The sparrow hawk (*Falco sparverius*) could even maintain body weight at an ambient temperature of 40°C for 48 h when it panted almost constantly. Bartholomew and Cade (1963) also pointed out that the large raptors often fly at high altitudes in air considerably cooler than at surface level.

One quantitative study of water turnover in a non-drinking carnivorous bird has been made by Duke et al. (1973) who fed great horned owls (*Bubo virginianus*) either laboratory white mice or day-old domestic turkey poults. The findings were essentially similar on the two diets, the only difference being that the poults contained a slightly higher fraction of water (67%) than the mice (62%), this resulting in a slightly higher water content of the excreta (81% versus 77%) and of the pellets (63% versus 61%). Calculated in per cent of body weight (average body weight 1.6 kg), the intake of water was 4.9% per day and the water loss in excreta 2.4%, in pellets 0.3%. This leaves 2.2% to be lost as evaporation, i.e. 45% of the total water intake. The observed values of water intake and rate of evaporation are thus not different from those observed in herbivorous birds on ad-libitum water intake (see Tables 2.1 and 4.1).

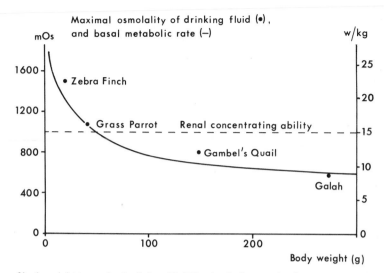

Fig. 2.3. Relation of body weight to maximal salinity of fluid intake. In four species that concentrate to about 1000 mOs, the maximal salinity (at minimal measurable drinking rate) is recorded as a function of body weight. The relation follows basal rate of metabolism. Reproduced from Skadhauge (1975)

2.4 Salinity Tolerance

As mentioned in the introduction (see p. 1), the maximal salinity of fluids that birds can ingest and still maintain body weight must be related to the maximal renal concentrating ability, since "free water" is generated by excreting NaCl at higher concentration than in the drinking fluid. The maximal concentration of the drinking solution is in reality identical to a NaCl ingestion rate as the real parameter that is recorded is concentration at minimal measurable drinking rate. To some extent the maximal salinity of the fluid ingested will be governed by the magnitude of the evaporative water loss and, for some species, the rate of excretion through the salt gland. If the rate of evaporation is relatively low, the amount of "free water" that the bird must procure is smaller, and the concentration of NaCl in drinking water closer to the renal ceiling. For small birds, which have a high rate of metabolism per kg body weight, there is a relatively high rate of production of metabolic water, which serves the same function and permits drinking of water of higher salinity. Skadhauge (1975) has shown that in four species that concentrate ureteral urine to about 1000 mOs, zebra finch, grass parrot (*Neophema bourkii*), Gambel's quail, and galah (*Cacatua roseicapilla*), the maximal salinity of water intake and the standard metabolic rate had the same relation to body weight (Fig. 2.3). The correlation rests on the assumption that the minimal measurable drinking rate is constant in relation to body weight. The relationship explains generally the intake of water of salinities up to the renal osmotic ceiling. The drinking of more concentrated solutions by the zebra finch requires other explanations (see p. 161).

Even with the limitations to the physiological interpretations of maximal salinity of drinking water this parameter gives an expression of the ratio of total salt

23

Table 2.4. Drinking of NaCl-solutions. The table presents the maximal salinity of the drinking solutions at which the birds maintain body weight. For some birds also indicated is the salinity at which drinking rate is maximal

Species	Notes	Maximal concentration of drinking fluid	fluid intake % of body weight/day	References
Zebra finch (*Poephila guttata*)	Max. intake at	0.3 *M* NaCl *0.8 M NaCl*	90% 3%	Skadhauge and Bradshaw (1974)
Savannah sparrow (*Passerculus sandwichensis beldingi*)		0.7 *M* NaCl	53%	Poulson and Bartholomew (1962a)
Savannah sparrow (*Passerculus sandwichensis brooksi*)		0.3 *M* NaCl	41%	
Black-throated sparrow (*Amphispiza bilineata*)	Max. intake at	0.2 *M* NaCl 0.4 *M* NaCl	30% 18%	Smyth and Bartholomew (1960b)
Tree sparrow (*Spizella arborea*)		0.15 *M* NaCl	39%	Ohmart and Smith (1970)
Brewer's sparrow (*Spizella breweri*)		0.50 *M* NaCl	29%	
Redwing blackbird (*Agelaius phoeniceus*)	Max. intake at	0.30 *M* NaCl 0.35 *M* NaCl	156% 56%	Hesse and Lustick (1977)
House finch (*Carpodacus mexicanus*)		0.3 *M* NaCl	110%	Poulson and Bartholomew (1962b)
Red crossbill (*Loxia curvirostra*)		0.2 *M* NaCl	100%	Dawson et al. (1965)
Purple grackle (*Quiscalus quiscula*)		0.2 *M* NaCl	153%	Harriman and Kare (1966b)
European starling (*Sturnus vulgaris*)		<0.2 *M* NaCl	—	
Sage sparrow (*Amphispiza belli nevadensis*)		<0.25 *M* NaCl	—	Moldenhauer and Wiens (1070)
Budgerigar (*Melopsittacus undulatus*)	Max. intake at	0.2 *M* NaCl	24%	Cade and Dybas (1962)

Table 2.4 (continued)

Species	Notes	Maximal concentration of drinking fluid	fluid intake % of body weight/day	References
California quail (*Lophortyx californicus*)		37.5% sea-water	—	Bartholomew and MacMillen (1961)
California quail (*Lophortyx californicus*)		0.30 *M* NaCl	—	Carey and Morton (1971)
Gambel's quail (*Lophortyx gambelii*)		0.45 *M* NaCl	—	
Vesper sparrow (*Pooecetes gramineus*)		0.25 *M* NaCl	39%	Ohmart and Smith (1971)
Mourning dove (*Zenaidura macroura*)		0.15 *M* NaCl	28%	Bartholomew and MacMillen (1960)
Ground dove (*Columba passerina*)		0.15 *M* NaCl	22%	Willoughby (1966)
Domestic duck	Max. intake at	0.14 *M* NaCl	14%	Fletcher and Holmes (1968)

excretion to ability to conserve water. It also indicates the extent to which birds without salt glands can drink water from salt lakes, estuary zones, and salt marshes. The interest in this information has prompted many investigators to offer dilutions of sea-water rather than solutions of pure NaCl (Table 2.4). Inspection of Table 2.4 shows good agreement between maximal salinity and renal concentrating ability, (see also Table 5.3).

2.4.1 Intake of Saline Solutions by Birds with Salt Glands

As mentioned previously (see p. 17), Harriman (1967) observed a reduction in drinking rate with increasing salinity in the laughing gull. Similar observations have been made by other workers. Schwarz (1965) injected herring gulls intraperitoneally with solutions of either fresh water or NaCl of 5%–9%. After the injection the birds were exposed to drinking solutions of either freshwater or NaCl solutions ranging from 1.5% to 9%. He observed a maximal intake of 1.5% of NaCl after freshwater injection and a marked reduction in intake of all saline solutions after the injection of 5%–9% of NaCl, as an inverse function of concentration. The birds thus select a higher intake of fresh water as a function of osmotic stress, all the injections being hyperosmotic to plasma. In the absence of osmotic stress 1.5% of NaCl would seem to be preferred. Fletcher and Holmes (1968) have addressed the problem of water intake in the domestic duck and measured drinking rates as functions of salinity from fresh water to 500 mM NaCl/l. A rate of freshwater drinking of 130 ml/kg day was observed. Intake was reduced to about 30 ml/kg day at 500 mM/l in such a way that above 150 mM/l (isosmotic fluid) the rate of intake of NaCl remained constant at approximately 20 mM/kg day. This constant intake demonstrates that these birds regulate salt intake and since the salt intake is considerably lower than the excretory capacity (see p. 141), they obviously avoid osmotic stress.

2.5 Salt Requirements of Domestic Birds

The studies described in the previous section have elucidated limitations to the concentration of salt in drinking water. This is mainly a problem of excretory ability in relation to evaporation, formation of metabolic water, etc. A salt preference or even salt craving is also apparent in some species although it is not easily quantifiable.

In this section investigations on requirements for Na, Cl, and K will be considered in relation to optimal growth rate, production of eggs, and general health. Minimal requirements, deficiency symptoms, and toxic levels have been investigated. Experiments have been performed on the domestic fowl, domestic turkey, the pheasant (*Phasianus colchicus*), and Japanese quail, but there are few reports on others. Many of the studies had the practical goal of defining nutritional requirements for production of domestic fowl. The present review is by no means exhaustive.

Halpin et al. (1936) surveyed the older literature on salt poisoning in poultry. In their own experiments optimal growth rate and highest egg production were found at 0.5% to 1.0% of salt. Higher salt concentrations produced watery droppings.The Na requirements of chicks have more recently been studied by Heuser (1952), by McWard and Scott (1961), and by Nott and Combs (1969). The first of these reports claimed 0.65% and 0.30% of salt, respectively, to be necessary for optimal growth rate, whereas the last found 0.11%– 0.13% of NaCl sufficient. This corresponds to 20 mM NaCl/kg food. The differing Na requirements as determined by various authors may be related to fibre content and other differences in the food. Minor differences in the NaCl content of drinking water may also be of importance (Ross 1979). A closer analysis is not possible.

Burns et al. (1953) examined the effects of Na, K, and Cl on the growth rate and mortality in chicks for experimental periods of up to 4 weeks. The requirements were Na: 0.10%–0.30%, K: 0.23%–0.40%, and Cl: less than 0.06%. Vogt et al. (1970) observed maximal gain in weight in chicks with a diet containing 0.12% of Na, a level also required by hens. The mortality of the chicks was significantly increased by 0.58% of Na or more.

The optimal NaCl requirement for egg production by the domestic fowl has received considerable attention. Burns et al. (1952) established a level of 0.19% of NaCl as minimal for maintenance of body weight and optimal egg production. The lowest NaCl level in a purified diet (0.04% of NaCl) resulted in a marked impairment of egg production. However, the few eggs laid displayed normal hatchability. Whitehead and Shannon (1974) observed that a diet containing 0.38% of Na, equivalent to an intake of 1.3 mequiv/Na per day, resulted in a cessation of egg-laying in hens with no other ill effects. After a period of 4 weeks on this diet reversion to normal dietary NaCl content resulted in resumption of normal egg production. On the low NaCl diet the birds behaved normally (Hughes and Whitehead 1974). Because the egg contains about 3 mequiv of Na the cessation of egg production may be inevitable when birds on the low intake of Na strive to maintain NaCl balance. Nesbeth et al. (1976), and Begin and Johnson (1976) have confirmed the cessation of egg production on a low Na diet. Both groups noted further a decrease in food intake. Begin and Johnson (1976) measured the amount of NaCl that must be added to a low salt diet to induce stable egg-laying. They found that 0.125% of NaCl reduced egg-laying by a third, whereas 0.25% of NaCl resulted in unchanged egg-laying.

Hurwitz and co-workers (Cohen et al. 1971; Hurwitz et al. 1973) have studied the Na/Cl ratio in the domestic fowl. In growing chicks they found maximal weight gain at a one-to-one Na and Cl content in the food, and the gain-to-feed ratio was maximal at contents of 0.13% of Na and 0.13% of Cl in the food. When the Na/Cl ratio was varied systematically in laying hens by addition of various salts of the two ions during shell formation, considerable changes were observed in plasma concentrations of Na, Cl, HCO_3, and pH as functions of the Na/Cl ratio.

The effect of isolated Cl deficiency was studied by Leach and Nesheim (1963). Lack of Cl resulted in extremely poor growth rate, high mortality rate, dehydration, haemoconcentration, and reduced Cl concentration of plasma. In addition, the chicks showed characteristic nervous symptoms. A diet of 34 mequiv Cl/kg food resulted in optimal growth rate and prevented the deficiency symptoms.

The minimum Na requirement and interaction of Na and K in the diet was tested in young domestic turkeys (0–4 weeks of age) by Kumpost and Sullivan (1966), who found that 0.15% to 0.20% of Na was necessary for maximal gain of body weight. The K requirement of turkey poults was studied further by Chavez and Kratzer (1973), who observed optimal growth rate and minimal mortality at a level of 0.60% to 0.75% of K, an equivalent of 15–19 mequiv/kg food. The minimum level of Cl for maximum growth is 0.15%, as measured by Kubicek and Sullivan (1973), who also observed deficiency symptoms similar to those in domestic fowl. For young pheasants and young Japanese quail a Na content of 0.09% and a Cl content of between 0.05% and 0.11% in food are sufficient for normal growth (Scott et al. 1960). Accordingly, they concluded that the diet should contain at least 0.15% of NaCl. In both species increased mortality was first noted at 7.5% of NaCl in the diet. Lunijarvi and Vohra (1976) measured growth rate weight of adrenal glands, and plasma Na in growing Japanese quail as functions of the Na concentration of a purified diet. All three parameters responded to an increased Na concentration of the diet and reached a stable level at 0.10%.

In conclusion, the NaCl requirements of several species of growing granivorous birds and of egg-laying fowl seem to be within 0.10% to 0.15% of NaCl in the food, equivalent to 17–25 mM/kg food, with slight changes due to other components of the diet. The observed "optimal growth rate" is probably associated with a maximal expansion of the extracellular volume.

A chronic high NaCl intake, by offering 0.9% NaCl as the only drinking solution, was observed to induce nephrosclerosis and oedema in 19-day old chicks, with birds dying within 20 days, 0.3% of NaCl was without ill effects (Seley 1943), 0.4% was, however, sufficient to cause weight reduction and increased mortality in poults and ducklings (Krista et al 1961). A NaCl intake over 3g/kg body weight resulted in cardiac hypertrophy, increased blood pressure, and hypertrophy of the renal glomeruli (Krahower and Heino 1947). A food content of 3% NaCl was found to be the minimal toxic level in growing chicks (Barlow et al. 1948).

Some investigators have examined the effect of either NaCl restriction or NaCl loading on body-water compartments. Restriction of dietary Na, induced in white leghorn pullets by a purified diet containing either 0.5% of NaCl or no NaCl addition at all (Harris and Koike 1977), resulted in an average daily excretion of 3.85 mequiv Na/day and 0.35 mequiv Na/day respectively. There was no difference in food intake, in plasma osmolality or plasma Na between the two groups, whereas extracellular fluid volume and Na_{ex} declined (see Sect. 1.3). The plasma volume and the haematocrit remained, however, unchanged. Domestic fowl thus respond rather efficiently to this tenfold reduction in NaCl intake as they are able to maintain a constant circulating blood volume. Pang et al. (1978) observed a 12% reduction when young poults were switched from a 0.25% to a 0.05% Na diet, and Abdel-Malek and Huston (1975) found a two- to threefold decrease of turnover rate for Na^{22}, depending on the temperature, when broiler chickens were switched to a low Na diet. After injection of hyperosmotic NaCl in domestic ducks, glaucous-winged gulls, and roosters Ruch and Hughes (1975) made estimates of total body water, extracellular fluid volume, and plasma volume (see Sect. 1.3). The domestic fowl had the smallest total body water and plasma volume; the gulls had the highest volumes in all compartments. After an intravenous injection of NaCl plasma Na increased by 20%. The sea-water-loaded domestic duck and the glaucous-winged gull responded with large increases in extracellular fluid volume of 26% and 33% respectively, whereas the freshwater-reared ducks and the rooster had lower increases of 11% and 17% respectively. Only the bird without a salt gland, the rooster, had a large increase in total body water. As only sea-water ducks and the gulls responded with a high rate of salt-gland secretion (see Fig. 7.4), the authors conclude that not only increase of plasma Na, but extracellular volume expansion is also necessary for full stimulation of secretion by the salt gland (see Chap. 7).

Chapter 3

Uptake Through the Gut

3.1 Absorption of Salt and Water in the Small Intestine

After salts, water, and nutrients have been ingested, they are normally absorbed in the small intestine together with water and salt from secretions from oesophagus, crop, proventriculus, gizzard, and glands of the digestive system. Investigations on the role of the processes of intestinal absorption and secretion for salts (Na, K, Cl) and water are reviewed in the first section of this chapter. In the second section the osmotic and solute composition of the intestinal contents will be considered. A third section considers the role of the intestinal caeca, when present, in osmoregulation.

Intestinal absorption and secretion of salts and water have been investigated by three techniques, (1) in-vivo installations, (2) in-vitro measurements with intestinal preparations, (3) collection, after sacrifice, of contents from various parts of the intestinal tract after feeding an unabsorbable water marker in the diet.

Duke et al. (1969) have studied absorption from ligated jejunal segments about 15 cm in length in vivo in Wrolstad medium white turkeys. They observed after 1 h isosmotic absorption of 75% of NaCl from instilled isosmotic NaCl solutions; Na concentration and osmolality in the segment remained constant. The Cl absorption rate was slightly higher, suggesting an anion exchange process, and a small secretion of K was observed. Mongin and deLaage (1977) perfused 19–20 cm of duodenum of laying Warren hens. Water flow was measured as change in concentration of an unabsorbable water marker, PEG-C^{14}. Without NaCl in perfusates of varying osmolality they observed a linear relation between net water flow and osmotic difference across the gut wall. With an impermeant solute, raffinose, in the lumen there was no net flow of water at a luminal osmolality of 158 mOs (the plasma osmolality was 322 mOs), suggesting a reflection coefficient of 0.5 (158/322) of the solutes of plasma. The reflection coefficient expresses the leakiness of the membranes to the solutes. It is unity in an ideal semipermeable membrane. In accordance with this low reflection coefficient (high solute permeability), a net inflow of Na and Cl into lumen occurs. F.M.A. McNabb (1969b) instilled 1% NaCl solutions into a section of the small intestine (from posterior to the pancreas to just anterior to the caeca) in three species of quail and observed an isomotic absorption of NaCl and water. Isosmotic absorption was also observed in vitro from everted gut sacs from five sections of the small intestine of domestic ducklings by Crocker and Holmes (1971), who found that the concentration of Na in the transported fluid is approximately the same as in the incubation medium.

The conclusion from these studies is that in the small intestine of birds, just as in mammals, net fluid movement is largely created by the Na (Cl) absorption. It is of interest whether there is any regulation of the transport of NaCl as a function of the oral salt intake. The study of Crocker and Holmes (1971) has also contributed to a resolution of this question in that they compared the Na absorption of ducklings on fresh water with those offered 60% of sea-water. After switching to sea-water the uptake of Na and water increased in parallel. It was highest in the anterior gut segments, reaching maximally twice the uptake on fresh water. The Na transport rate was also measured as a function of length of time that the birds were exposed to sea-water. The transport rate was already high after 1 day and maximal after 48 h; it had not declined 1 day after switching back to fresh water, but decreased to basal level on the third day. Since spironolactone prevented the sea-water response, this homeostatic regulation is most likely mediated through an adrenocortical hormone that facilitates sea-water adaptation in marine birds (see p. 154).

Opposite responses to salt loading and depletion were observed by a different technique in the domestic fowl by Hurwitz et al. (1970) in white leghorn laying hens fed high and low Na diets. An unabsorbable water marker, Yttrium-91, was fed for 4 days in the diet. The hens were then sacrificed and the concentrations of water marker and Na and K determined in five segments of the gut. Na depletion reduced the Na concentration of the chymes delivered to the colon, and an increase in K concentration was observed. The Na/K ratio changed from 1.4 on the high to 0.5 on the low Na diet, suggesting mediation via an adrenocortical hormone. The Na/Yt-91 ratios indicated a considerable secretion of Na into crop and duodenum, with a rapid absorption in jejunum and ileum. When birds on high and low Na diet were compared Na absorption was higher in those on high Na diet in the jejunum, but lower in the ileum. But the overall fractional absorption was similar, 41% in those on high Na diet and 45% in those on low Na diet. Due to the larger intake in high-Na-diet birds, the absolute Na absorption rate was more than twice as high compared with birds on low Na diet. The transmural electrical potential difference (PD) was not measured in this study, but as PD values from gut lumen to blood of only 6–12 mV lumen negative have been reported for the domestic fowl (Hurwitz and Bar 1969), the concentration difference will largely determine the driving force for transepithelial ion flow.

3.2 Osmotic and Solute Composition of Contents of the Small Intestine

Observations are available almost exclusively on the domestic fowl, domestic duck, and domestic goose. Most studies have been carried out to solve problems unrelated to osmoregulation, such as the environment of intestinal parasites or correlation of type of food to intestinal amino acid composition. Reported values usually have a wide range, as is to be expected, since they vary not only as functions of type of food and electrolyte intake, but are also related to time after last eating, phase of egg laying, level of hydration, and time of day. Available findings are summarised in Table 3.1. When obtaining material for examination several authors have distinguished between an upper and a lower jejunum and ileum. For simplicity Table 3.1 includes only information from duodenum and aboral ends of jejunum and ileum. It appears that the duodenum contains a hyperosmotic aqueous solution with somewhat low concentrations of strong electrolytes. The high osmolality is presumably due to rapid hydrolysis of carbohydrates and proteins. Further caudally the intestinal contents are only slightly hyperosmotic to plasma with the osmolality dominated by the strong electrolytes, Na, K, and Cl, but with significant concentrations of glucose, amino acids, and volatile fatty acids (VFA). VFA are largely produced as a result of microbial fermentation as they are nearly absent in germ-free animals (Annison et al. 1968). Much higher concentrations of VFA were observed in the caeca (see p. 38). It is noteworthy that the lower ileum contains fairly high concentrations of glucose and amino acids. In the colon such chyme will allow stimulation of the Na absorption of hens on a high NaCl diet as observed in vitro (see p. 120). The water content of ileum (85%–80%) is not greatly different from that of the defecated mass of urine and faeces of birds with free access to water (see p. ß). This serves to illustrate the efficient water conservation of kidney and cloaca.

3.3 Caecum: Role in Osmoregulation

The role that the intestinal caeca play in conservation of salt and water can be elucidated in various ways. A basic approach is comparative. Which species have caeca? Is there any correlation between salt-water relations (high or low salt intake, desert adaptation, etc.) and development of the caeca among species, families, and orders? Is there any relation among individuals within species between the type and degree of osmotic stress and development of the caeca? Another approach is through physiological experimentation. What is the general function of the caeca? Is the osmoregulation impaired if they are extirpated? Are they only functional during osmotic stress? Is intestinal water absorbed in the caeca? Does ureteral urine pass retrogradely as far as the caeca? If so, which fraction, and are components of the urine resorbed? What is the fate of solutes and molecules in suspension? Are uric acid and urates present in the caeca, and are they metabolised by bacteria? What is the solute transport capacity of the epithelium? Some of these questions can now be answered, at least partially, but others not at all.

3.3.1 Presence of Caeca

The occurrence of caeca among families and orders of birds has been surveyed thoroughly. Their anatomy was meticulously described in three publications just after the turn of the century (Mitchell 1901; Maumus 1902; Magnan 1911). The anatomy, as related to nutrition, has been reviewed by Ziswiler and Farner (1972). No consistent plan with respect to osmoregulation emerges from these publications. Maumus (1902) concluded that birds with rudimentary caeca were predominantly carnivores, and those with developed caeca predominantly herbivores. The presence of caeca is, however, in this context far from obligatory.

Table 3.1. Solute concentrations of contents of the small intestine. Mean values ($\pm$ S.E.)

Substance	Species	Duodenum	Jejunum	Ileum	References
Osmolality mOs	Domestic fowl	572±21	573±20	451±13	Mongin (1976a)
	Domestic fowl	—	—	424±11	Skadhauge (1968b)
	Domestic duck	—	—	324±2	Crompton and Edmonds (1969)
	Jackass penguin[b] (*Sphenicus demersus*)	614	448	445	Erasmus (1978b)
Water %	Domestic fowl	86	86	81	Mongin (1976a)
	Domestic duck	93	85	85	Crompton and Nesheim (1970)
Na mequiv/l	Domestic fowl	66±3	95±3	116±5	Mongin (1976a)
	Domestic fowl	—	—	105±11	Skadhauge (1968b)
	Domestic fowl	74±6	59±8	112±13	Hurwitz et al. (1970)
	Jackass penguin[b]	238	162	142	Erasmus (1978b)
K mequiv/l	Domestic fowl	47±2	35±2	36±3	Mongin (1976a)
	Domestic fowl	—	—	61±15	Skadhauge (1968b)
	Domestic fowl	36±2	21±2	80±12	Hurwitz et al. (1970)
	Jackass penguin[b]	17	11	11	Erasmus (1978b)
Cl mequiv/l	Domestic fowl	95±4	95±3	14±3	Mongin (1976a)
	Domestic fowl	—	—	125±17	Skadhauge (1968b)
	Jackass penguin[b]	268	162	193	Erasmus (1978b)
VFA mM	Domestic fowl	8±2	10±2	—	Annison et al. (1968)
	Domestic goose	—	4±2	5±2	Clemens et al. (1975)
Glucose mM	Domestic fowl	—	—	32	Crompton (1966)
Amino acids mM[a]	Domestic fowl	—	—	2	Crompton (1966)
	Domestic fowl	—	32	—	Bielorai et al. (1972)
	Domestic fowl	10	17	8	Crompton (1966)

[a] Based on an average molecular weight of 120
[b] 35 min after an oral salt load

33

Caeca are, for example, totally absent in parrots. The observations of Maumus should be reexamined in light of modern concepts of evolution. Magnan (1911) confirmed the conclusions of Maumus, and stated body weight, caecum length, and distance (along the colon) from caecum to the rectal ampulla. Most impressive was a caecal length of 84 cm in a *Tetrao urogallus* of 4.1 kg with a colon length of 26 cm. Also, in the *Rhea americana*, the caecum is large, 57 cm in a 10.6 kg bird, with a colon length of 34 cm. In these birds not only the caeca, but also the colon would seem to have large areas. As the function of the caeca is related to microbial fermentation and production of VFA (see p. 38), it is noteworthy that the caeca are particularly well developed in the Galliformes. Browsing species (grouse, ptarmigan, and capercailzie) have longer caeca than grain-eating species (quail, partridges, pheasants, and turkeys; Leopold 1953). Gasaway (1976a) observed an average caecum length of 104 cm in the willow ptarmigan (*Lagopus lagopus*) (average body weight 548 g). In the domestic fowl caeca are also fairly large, often to a length of 20 cm with a maximal diameter of 1–1.5 cm and filled with yellow-brown faeces, very different from cloacal faeces. Since the caecal faeces are easily distinguishable from cloacal faeces, separate collection of them for physiological experiments is relatively simple. Since emptying of the caeca proceeds rapidly and just after emptying of the cloaca, mixing of cloacal and caecal faeces should be limited (Fenna and Boag 1974a). The caeca are evacuated once every 24–48 h, i.e. once for every seven to eight cloacal discharges (Röseler 1929). Duke et al. (1968) observed the same ratio between cloacal and caecal defecation in the ring-necked pheasant. Using single-dose feeding of Cr^{51} they observed a four-fold faster passage time of cloacal than of caecal samples. Gasaway et al. (1975) estimated that 59% of the caecal contents are discharged per caecal defecation.

3.3.2 Flow of Material into the Caeca

Fluids may gain entrance to the caeca from two directions, either from the ileum or by retrograde movement from the colon. The colonic inflow may contain both ureteral urine, as modified during the passage through coprodeum and colon, and fluids originating from the small intestine. Studies by Browne (1922), and by Olson and Mann (1935) on the domestic fowl indicate that only liquid chyme gains entrance to the caeca. Only water-soluble dyes fed to the birds were recovered in the caeca, whereas particulate matter such as lamp black bypassed the caeca; colloidal carbon, however, was taken in. These results are in agreement with those obtained by Gasaway et al. (1976) and by Clemens et al. (1975).

Retrograde flow of fluids from the coprodeum to the very end of the caeca was demonstrated by Browne (1922) and by Skadhauge (1968) by instillation of fluids containing Evans blue or methylene blue into the cloaca through the anus. The dye was visible when the caeca were opened after sacrifice. Skadhauge (1968) even found tracks of uric acid in the narrow neck of the caecum close to the ileum. Bell and Bird (1966) failed to detect uric acid in the bulk of caecal material near the blind end. Since most urinary uric acid is precipitated in the coprodeum, and little moves to the colon (Skadhauge 1968), and the amount in solution is small (see p. 85), it is not surprising that no uric acid is measurable in the caeca. In addition there is the

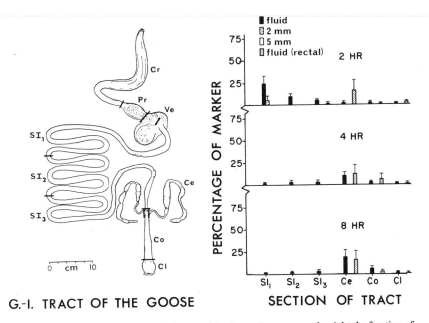

G.-I. TRACT OF THE GOOSE   SECTION OF TRACT

Fig. 3.1 On the left is shown the gastrointestinal tract of the domestic goose; on the right the fraction of various markers recovered in the contents of different segments of the gut (see text). *Cr* Crop, *Pr* proventriculus, *Ve* Gizzard, *SI1-3* small intestine, *Ce* caecum, *Co* colon, *Cl* coprodeum. The most significant findings is that nearly 20% of both orally and rectally instilled fluids are recovered from the caecum. Reproduced with permission from Clemens et al. (1975)

possibility of microbial metabolism of uric acid. Radiographic evidence in the domestic fowl (Akester et al. 1967; Koike and McFarland 1966) and in the roadrunner (Ohmart et al. 1970), proves that ureteral urine is transported from colon (see p. 97) to the blind ends of the caeca. Vitamins (Polin et al. 1967) and urea (Bell and Bird 1966) of urinary origin have also been demonstrated in the caeca.

A quantitative estimate of the fraction of the ureteral urine that reaches the caeca was not made in any of these studies. Skadhauge (1973; p. 48) found in the faeces coming from the caeca of dehydrated domestic fowl a fraction of up to 20% of the total amount of polyethylene glycol 4000 (PEG) recovered in the droppings after intramuscular injection into the breast muscle of this unabsorbable water marker, which is excreted by filtration in the kidney (see p. 60), and a valid marker of the intestinal aqueous water phase (Mongin 1976b).

Virtually the same fraction of water marker was found in the caeca of geese by Clemens et al. (1975), who presented geese simultaneously with a PEG-containing solution orally and with 2 mm and 5 mm markers of particulate matter. Cr[51]-EDTA was instilled in the cloaca. A number of birds were killed after 2, 4, 6, 8, and 12 h, and the fractions of the instilled substances present in various segments of the gut were recorded (Fig. 3.1). The markers of particulate matter were retained for more than 12 h in the gizzard. A significant fraction, nearly 20%, of the oral fluid marker was found in the caeca after 8 h. The cloacal (rectal) fluid marker had accumulated after 2 h in the caeca, retrograde flow having reached a peak fraction

35

of 18% after 8 h. It can therefore be concluded that an important fraction of a cloacal water marker gains access to the caeca as does a similar fraction of an orally ingested water marker.

X-ray studies by Fenna and Boag (1974a) on the Japanese quail indicate a rapid passage of the water marker through the small intestine into the cloaca from which it is then regurgitated into the caeca. Similar observations have been made in the domestic fowl by Helnan (1949). The findings of Clemens et al. (1975) also suggest such a regurgitation.

3.3.3 Mechanical Regulation of Inflow in the Caeca

The driving force for the flow of fluids into the caeca is known to some extent: X-ray studies with $BaSO_4$ in the cloaca have clearly demonstrated antiperistaltic waves in the domestic fowl (Akester et al. 1967), in the domestic turkey (Lai and Duke 1978), and the Japanese quail (Fenna and Boag 1974a). Antiperistaltic waves were also observed directly in the laying domestic fowl through a window implanted in the abdominal wall (Yasukawa 1959). In this way fluid around the central faeces core is squeezed orally and "sucked" into the caecum. Fenna and Boag (1974a) observed that "peristaltic and antiperistaltic waves of contraction converging at the openings of the caeca apparently forced intestinal contents laterally into the caeca". The pressure in the lower ileum during peristaltic waves has been measured to be 16.2 Torr (Duke et al. 1975), and correlation of radiographic observations and recordings of electrical activity have shown small waves responsible for the antiperistaltic flow and large waves for the forward movement of digesta (Lai and Duke 1978). These authors suggest that "urine is moved along the mucosal surface by antiperistalsis and colonic ingesta are moved through the center of the lumen by the large contractions".

None of these studies have, however, been repeated in dehydrated birds in which water retention is likely to be pronounced. Skadhauge (1968) observed that "dehydrated birds would keep the anus closed and retain instilled fluids or gels for very long periods whereas these substances would be quickly expelled by normally hydrated birds". This observation suggests a regulation of the extent to which urine is forced orally in the intestine.

3.3.4 Quantitative Studies of Caecal Function

Gasaway et al. (1976) have effected a meticulous investigation of digestion of dry matter and absorption of water in the intestine and caeca of rock ptarmigans (*Lagopus mutus*). The birds were fed two isotope markers, Cerium[144] to determine digestibility of particulate matter, and Cr^{51}-EDTA, which served as an unabsorbable water marker. The authors measured an intestinal dry-matter output of 9.6 g/day and a water excretion of 23.2 ml/day, including ureteral urine. The values for caecal droppings were 1.7 g dry matter/day and 5.9 ml water/day. Only 14% of the Cr^{51} was excreted in the cloacal droppings, but 86% in the caecal droppings. Assuming no dry matter absorbed in the colon, the authors calculated a 12% absorption of water in the colon from the water fraction in the contents in its upper and lower end. In the droppings the marker of particulate matter was distributed with 82% in the cloacal droppings and only 18% in the caecal droppings. One can therefore conclude that a large fraction of water from the small intestine enters into and is resorbed in the caeca, whereas these organs play a minor role in the digestion of the highly digestible food, which these birds received in captivity.

From the concentrations of Cr^{51} in the water phase of caecal and cloacal droppings and the total rate of excretion, and absorption of water in the large intestine, the authors calculated the absorption of water in caeca to be 163 ml/day or 384 ml/kg 24 h. This indicates that 98% of all water absorbed from the hind gut occurred from the caeca. The calculation can, however, be accepted only with some reservation. First, such a high caecal water absorption in the absence of a marked disappearance of dry matter would require a high water content of the chyme of the terminal ileum. If the dry matter of the excreta totals 11.3 g/day and 0.3 is absorbed in the caeca, even a water fraction as high as 90% of ileal contents would result only in a delivery of 104 g H_2O/day to the lower small intestine. A very high value of 94% of H_2O would be necessary to account for the calculated caecal absorption. Such a value for ileal water content is far beyond the water content measured in other birds (see Table 3.1). Second, the cloacal faeces would be diluted by ureteral urine. This would give a Cr^{51} concentration that is too low, thus leading to overestimation of the caecal absorption rate. Since the cloacal droppings, however, had a water fraction of only 70.7%, evaporation may have occured. The reason for the high caecal value must consequently be found elsewhere. One possibility is that the high fraction of Cr^{51} enters the caeca in a preferential inflow bound to very particulate matter. Furthermore, a regurgitation of fluid of ileal origin from colon to the caeca (see p. 97) would complicate the calculation. Finally, the small amount of Cr^{51} absorbed from the small intestine (Gasaway et al. 1975) would be excreted in urine with a high urine to plasma ratio. Be that as it may, the study has elucidated the significant physiological mechanisms by which the caeca conserve water and concentrate fine particulate matter.

As in the domestic fowl, the findings of Gasaway et al. on ptarmigan indicate a clear separation of particulate material, which bypasses the caeca, and liquid material, which accumulates in these organs. A similar conclusion was reached by Fenna and Boag (1974b) who observed on Japanese quail and spruce grouse (*Canachites canadensis*) that bulky, largely undigestible, cellulose material was rapidly discarded from the intestine and nutrient-rich residue retained in the caeca for longer periods of time. Inman and Ringer (1973) also found evidence of differential inflow of liquid material into caeca of Chukar partridge, (*Alectoris graeca*), ruffed grouse (*Bonasa umbellus*) and bobwhite quail which should permit further digestion and absorption.

The absorption of VFA, glucose, and water from the caeca was studied in vivo and in vitro in the domestic fowl by Parhon and Bârză (1967). These authors measured an absorption after 2 h of 66% of the glucose instilled in the caecum in vivo in a Tyrode solution with 3% of glucose; in vitro 26% was absorbed after 2 h. Water was only absorbed from solutions which contained glucose. The absorption of water after 1 h was 45% of the amount instilled, and 60% after 2 h. If according to these results a resorption rate of 1.2 ml/h per bird for the two caeca is assumed, an amount of water equal to the rate of urine flow in dehydrated fowls, about 5% of the body weight per day, can be absorbed by the caeca. Concerning absorption of VFA Timet et al. (1971) observed much more rapid passage of labelled acetate and propionate in the mucosa-serosa direction than vice versa.

As precise quantitative measurements of caecal water absorption are not available, considerable interest is attached to the consequence of ligation or extirpation of the caeca. These organs are neither necessary for maintenance of adult life, nor for normal growth if removed soon after hatching. Thus, Sunde et al. (1950) found equal body weight and egg production in normal and caecectomised domestic fowl. Beattie and Shrimpton (1958) caecectomised newly hatched cockerels and found, after 12 weeks, that body weight was not significantly different from that of controls. A similar observation was made for the domestic turkey by Schlotthauer et al. (1933). In none of these studies, however, were the birds challenged with dehydration or feeding of a low-quality diet. These procedures might have revealed the survival value of caeca. As it was to be expected (see p. 39), caecectomy is followed by a marked reduction in the cloacal excretion of VFA (Annison et al. 1968).

3.3.5 Faecal Water Content After Caecectomy

In three investigations authors have reported a higher water content in faeces after caecectomy. The estimates are semi-quantitative, at the best, as no special precautions were taken to avoid evaporation and only a few subjects were used. Radeff (1928) compared one operated domestic fowl to two controls. He observed "moister" faeces in the bird without caeca. Röseler (1929) had two operated hens on a diet of wheat from which faecal (including urinary) dry matter of 13%–17% and 7%–16% were observed, compared with 22% for normal birds. Thornburn and Willcox (1965) report average values of two cockerels before and after caecectomy. They received four types of food (wheat, pellets, barley, and oats). After the operation the average percentage of dry matter was reduced in one bird on three types of food, in the other on two. In the other experiments the percentage of dry matter was actually increased. None of the authors exposed their birds to dehydration.

The available evidence seems far from conclusive, although a role of the caeca in water conservation in the domestic fowl is certainly suggested. As caecectomy also leads to a change in digestibility of dry matter and because possible changes in renal water handling must be controlled, simple collection of faeces under oil will not resolve the problem. A total water budget in normal and operated birds on the same diet must be worked out.

3.3.6 Role of Caecum in Avian Nutrition

The predominant role of the caecum seems to be production of VFA by microbial fermentation. Sunde et al. (1950) found the caeca to be the site of greatest concentration of intestinal microorganisms in the domestic fowl, and Barnes (1972) reported a concentration of 10^8–10^{10}/g wet weight of uric acid utilising anaerobic bacteria in caeca of several domestic species. These bacteria were further characterised by Barnes and Impey (1974). Gasaway (1976a, b) measured a high

Table 3.2. Composition of caecal contents. Mean values ($\pm$S.E.)

Species	Substance	Value	Unit	References
Domestic fowl	H_2O	75	%	Mongin (1976a)
Domestic fowl dehydrated	Osmolality	491 ± 22	mOs	Skadhauge (1968)
	Na	30 ± 9	mequiv/l	
	Cl	53 ± 9	mequiv/l	
	K	73 ± 22	mequiv/l	
Domestic fowl	Urea	3 ± 1	mM/kg wet weight	Bell and Bird (1966)
	VFA	76 ± 9	mM/kg wet weight	
Domestic fowl	VFA	107 ± 4	mM/kg wet weight	Annison et al. (1968)
Domestic goose	VFA	38 ± 8	mM/kg wet weight	Clemens et al. (1975)
Rock ptarmigan (*Lagopus mutus*)	VFA	42 ± 13	mM/kg wet weight	Gasaway (1976b)

ratio of VFA production in the willow ptarmigan and in the rock ptarmigan, responsible for 11%–18% of basal metabolic rate. Barnes (1972) suggested that caecal size might be related to the need to supplement a poor diet, since recycling and metabolism of nitrogen from urinary uric acid might aid in nitrogen metabolism. Mortensen and Tindall (1978) found a marked uricase activity in the caecum of grouse (*Lagopus*). They suggested that "there is a rapid and effective breakdown of uric acid in the caecum and by this means the microorganisms may allow the host-bird to re-utilise the nitrogen moiety of the excreted uric acid". The quantitative role of this process should be investigated. Other investigators have suggested that an important function of the caecum is to metabolise crude fibre (Röseler 1929; Radeff 1928; Inman 1973).

The measured solute concentrations of caecal contents are summarised in Table 3.2. By comparison with Table 3.1 it will appear that the Na concentration in the caeca of the domestic fowl is much lower than in ileum and colon. This may reflect a high rate of Na-acetate absorption in caeca as observed in colon of ruminating mammals, particularly the goat (Argenzio et al. 1975). The concentration of Cl also seemed lower (see Tables 3.1 and 2), an observation that was also made when Cl was measured as a fraction of dry matter (Tyler 1946). Table 3.2 shows a high concentration of VFA in the caecal contents. Approximately two-thirds of the total concentration is acetate, with propionate next, around 20%, as studied in the willow ptarmigan (McBee and West 1969) and in the domestic duck (Miller 1976).

Summary. The main function of caeca in avian nutrition seems to be the digestion of fine particulate matter, crude fibre, and the production of VFA. The possibility of saving of nitrogen by bacterial breakdown of uric acid has been suggested. The caeca of several species may play an important role in osmoregulation as they receive around 20% of an unabsorbable water marker both from ureteral urine and after oral ingestion. Available studies indicate a caecal water absorption of 5%–38% of body weight/day, but caecectomy in the domestic fowl was not followed by marked changes in water output. Quantitative perfusion studies should be made.

39

Chapter 4

Evaporation

Whereas excretion of water through kidney, cloaca, and salt gland is highly regulated, evaporation of water is usually considered as an unavoidable "loss". But evaporation serves also an important function in thermoregulation since dissipation of heat, in birds as in other terrestrial animals, is partly achieved by evaporative cooling. When ambient temperature nears the temperature of the body surface, heat loss by radiation, conduction, and convection falls towards zero, and evaporation of water is the only mechanism for regulating body temperature. As sweat glands are absent from birds, evaporation from the respiratory system assumes a great role and orderly relations between oxygen uptake, evaporation, and water intake are, therefore, likely to be found. The relationship between oxygen uptake and rate of evaporation may, however, be modified by presence of evaporative surfaces in outer airways and air sacks beyond the regions of respiratory gas exchange. These allow panting and gular flutter, which dissociate water loss from oxygen uptake.

In this chapter evaporation is treated largely from a quantitative viewpoint, first, in relation to body weight and oxygen uptake for resting birds without thermal stress, second, for birds under heat loads, and third, as modified during simultaneous dehydration. Evaporation during flight will be referred to briefly in Sect. 10.2. Thermoregulation as such has been studied extensively in birds. The subject, beyond what is strictly related to osmoregulation, will be covered here only by reference to recent reviews.

4.1 Evaporation from Resting Birds Under Thermoneutrality

4.1.1 Relationship Between Body Weight and Oxygen Uptake

In a classical paper Bartholomew and Dawson (1953) measured evaporative water loss in 12 species of birds of the arid zones of the south-western United States, in which they observed an inverse relationship between body weight and evaporative water loss. Body weights of these species ranged from 10 to 150 g. Two other parameters are functions of the body weight. The first is standard metabolic rate or rate of O_2 uptake, the second is rate of production of metabolic water. O_2 uptake is related to water balance because respiratory gas exchange involves an evaporative water loss, as is also rate of production of metabolic water because it counteracts water loss. Metabolic water production is for a given ratio of metabolism of protein, fat, and carbohydrate also proportional to oxygen uptake. As dissipation of heat is achieved partly by evaporation; the fractional heat loss dissipated by evaporative cooling can be calculated. The total heat production is proportional to the O_2 uptake.

The section deals first with the basic relations between the above-mentioned parameters, and second with comparisons among various species.

The standard metabolic rate (SMR) in W [one W (Watt) equals 0.860 Kcal/h.] is in nearly all animals a power function of the body weight (BW) in kg: SMR $= a \cdot BW^b$, which is equal to log SMR $= b \log BW + \log a$.

In a double logarithmic plot the relation becomes linear. Lasiewski and Dawson (1967) reviewed previous findings and summarised measurements in a large number of birds ranging in size from humming birds (3 g) to the ostrich (*Struthio camelus*) (100 kg). They concluded that one straight line (n = 72) characterised all non-passerine birds: SMR $= 3.79 \cdot BW^{0.723}$.

The following equation characterised the investigated passerine birds (n = 42): SMR $= 6.25 \cdot BW^{0.724}$.

The regression line for all observed birds was

$$\text{SMR (W)} = 4.19 \, BW \, (kg)^{0.668} \tag{1}$$

and instinguishable from that observed for mammals. A similar regression line was made by Aschoff and Pohl (1970). The standard metabolic rate for resting birds under thermoneutrality is thus well established. The values of the function are such that O_2 uptake for birds weighing more than 60 g is falling slowly with increasing body weight, but below 60 g O_2 uptake raises steeply with decreasing body weight.

The relation between numbers of joules (One kJ is equal to 0.239 kcal) produced and litres of O_2 taken up is around 20 kJ/l O_2 (STPD) depending on the ratio between metabolism of carbohydrates, fat and protein. The metabolic water production is about 0.6 g H_2O/g carbohydrate, 1.1 g H_2O/g fat, and 0.5 g H_2O/g protein with uric acid as the end product. For a metabolic ratio among carbohydrate, fat, and protein of 1.00:0.07: 0.18 for seed-eaters (Krag and Skadhauge 1972) the metabolic water production is equal to 0.68 g H_2O/l O_2. This is thus the maximal rate, which evaporative water loss can assume without surpassing the rate of production of metabolic water. Some xerophilic birds actually evaporate less (Table 10.1).

This rate of metabolic water production is equal to 0.63 g H_2O/kJ. Use of the equation relating standard metabolic rate to body weight predicts a metabolic water production of 7.0% of body weight for a 5 g bird, 5.6% for a 10 g bird, but only 3.3% for a 50 g bird and 2.3% for a bird weighing 150 g. Similar values were calculated by Bartholomew and Cade (1963; Fig. 4.1). It is quite obvious that in order for xerophilic birds to remain in water balance without the intake of free water (see Table 2.2) the rate of evaporation must not surpass the metabolic water production, and the cloacal water loss not surpass the amount of preformed water in food. These conditions are fulfilled in some xerophilic birds (see Chap. 10).

Dissipation of heat by evaporation of water gives about 2.4 kJ/g H_2O with small variations depending on body surface temperature. The metabolic water production can thus generate a maximal heat dissipation of $2.4 \cdot 0.68 = 1.6$ kJ per 20 kJ produced or only 8% of the heat production. It will thus be seen that for evaporation to assume a major role in heat dissipation (when ambient temperature

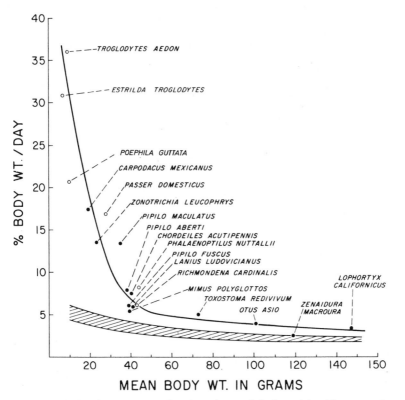

Fig. 4.1. Relation between rate of evaporation and body weight. The rate of evaporation at thermoneutral ambient temperature is expressed as percentage of body weight/day. The *lines shaded curve* indicates the calculated rate of production of metabolic water. The rate of evaporation increases more rapidly for body weights under 40 g. This results in a high drinking rate (see Table 2.1). Reproduced with permission from Bartholomew and Cade (1963)

approaches body temperature) external water must be available; if not birds become dehydrated. When respiratory evaporation proceeds from closed air passages, the maximal rate of evaporation is given by the water content of saturated air at body temperature after subtraction of the water content of the air, which is inspired. Calculations of this maximal rate of heat dissipation as related to ventilation can, be of quantitative value if the exact temperature of expired air and the fraction of evaporation occuring from the body surface both are known.

4.1.2 Relationship Between Body Weight and Evaporation

Since the pioneer work of Bartholomew and Dawson (1953) (included in Fig. 4.1) numerous other publications have confirmed the inverse relationship between body weight and rate of evaporation under thermoneutral conditions. Crawford and Lasiewski (1968) summarised the findings in 42 species ranging from humming-birds and finches (weights 3.0–18 g) to the rhea, emu, and ostrich (maximal weights

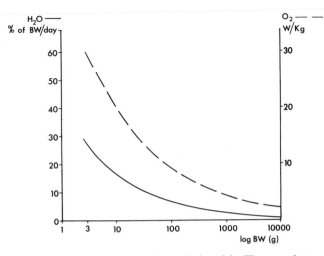

Fig. 4.2. Evaporative water loss and oxygen consumption in relation to body weight. The curves have been drawn according to equations (1) and (2), see text. The curves represent regression lines for values of all species investigated. Both parameters are expressed as functions of body weight. The rate of evaporation is reduced to fractionally lower values than the O_2 consumption at increased body weight. This reduces the amount of water evaporated per litre O_2 thus conserving water

20–100 kg). The following equation describes for all species included the relation between evaporation (E) and body weight (BW) (n = 53):

$$E \ (\text{g } H_2O/\text{day}) = 0.432 \ \text{BW g}^{0.585} \tag{2}$$

. The rates of evaporation and the oxygen consumption, according to equations (1) and (2) are shown in Fig 4.2.

Although the authors rightly point to some variation in experimental conditions such as differing relative humidity, state of nutrition, relation to meals, and hour of the day, the overall agreement is impressive, and the predictive value good. Dawson (1965) noted that even among four xerophilic Australian parrots the agreement between the actual measurements and the prediction of the equation was no better than for all observed species of birds. Concerning experimental techniques Lasiewski et al. (1966a, b) observed the importance of precise estimates of relative humidity during open flow measurements of evaporation. These authors found, however, good agreement between earlier direct weighing measurements and the open flow measurements. Crawford and Lasiewski (1968) pointed out that the exponent of body weight was lower for evaporation than previously measured for oxygen consumption (see p. 41 and Fig. 4.2). This implies that the amount of H_2O evaporated per l O consumed falls with increasing body weight. The value is 2.5 g H_2O/l O_2 for a body weight of 10 g (the typical weight of a finch), but it is reduced to 1.0 g H_2O/l O_2 for a bird weighing 38.3 kg (e.g. the emu). Naturally, in the abundant simultaneous measurements of rate of evaporation and oxygen consumption values between 1 and 2 g H_2O/l O_2 have been observed. When these values are compared to the metabolic water production (see also Fig. 4.1) it is

Table 4.1. Rate of evaporation. Evaporation is related to body weight at thermoneutrality, and during maximal thermal stress. In the latter situation is indicated the fraction of heat production which can be dissipated by evaporation. The maximal ambient temperature within the zone of thermoneutrality is also indicated

Species	Notes	Evaporation at thermo-neutrality % of BW/day	Maximal rate of evaporation % of BW/day	Fraction of heat dissipation %	Highest temperature at thermo-neutrality °C	Body weight (g)	References
Silver-bill (*Lonchura malabarica*)	Dehyd.	17	55	90	36	10	Willoughby (1969)
		12	22	77	36	—	
White-fronted chat (*Epthianura albifrons*)		18	47	56	36	10	Williams and Main (1976)
Crimson chat (*Epthianura tricolor*)		15	36	43	37	10	
Orange chat (*Epthianura aurifrons*)		15	32	37	37	10	
Stark's lark (*Spizocoryx starki*)		13	—	—	—	18	Willoughby (1968)
Grey-backed finch-lark (*Eremopterix verticalis*)		10	—	—	—	18	
Sage sparrow (*Amphispiza belli nevadensis*)		16	61	110	33	19	Moldenhauer (1970)
Cowbird (*Molothrus ater*)	Minimal water	22	—	78	—	33	Lustick (1970)
		3.4	42	62	40	33	
Ring-billed gull (*Larus delawarensis*)	Hatchling birds	10	67	150	35	35	Dawson et al. (1976)
Cassin's finch (*Carpodacus cassinii*)		11.9	134	208	38	28	Weathers et al. (1980)
Budgerigar (*Melopsittacus undulatus*)		5.6	89	156	33	31	Weathers and Schoenbaechler (1976)
Inca dove (*Scardafella inca*)	Nocturnal	7.2	38	107	—	42	MacMillen and Trost (1967a)
		2.5	—	—	—	42	MacMillen and Trost (1967b)
Munk parakeet (*Myiopsitta monachus*)		5.4	65	153	34	79	Weathers and Caccamise (1975)

Table 4.1. (continued)

Species	Notes	Evaporation at thermo-neutrality % of BW/day	Maximal rate of evaporation % of BW/day	Fraction of heat dissipation %	Highest temperature at thermo-neutrality °C	Body weight (g)	References
Whiskered owl (*Otus trichopsis*)	Nocturnal	2.8	—	—	30	121	Ligon (1969)
Elf owl (*Micrathene whitneyi*)	—	3.6	—	—	30	46	
Western plumed pigeon (*Lophophaps ferruginae*)		4	54	200	35	81	Dawson and Bennett (1973)
Spotted nightjar (*Eurostopodus guttatus*)		7	84	250	35	90	Dawson and Fisher (1969)
Cattle egrets (*Bubulcus ibis*)	1 day 10–12 days	—	69 65	140 200	— —	25 134	Hudson et al. (1974)
Sooty tern (*Sterna fuscata*)		8	50	89	36	148	MacMillen et al. (1977)
Bobwhite (*Colinus virginianus*)		3.1	—	—	—	181	McNabb (1969a)
California quail (*Lophortyx californicus*)		3.5	—	—	—	149	
Gambel's quail (*Lophortyx gambelii*)		3.2	—	—	—	149	
Roadrunner (*Geococcyx californicus*)		4.0	29	136	—	285	Calder and Schmidt-Nielsen (1967)
Domestic pigeon		4.1	31	125	—	359	Marder (1973)
Domestic fowl (Beduin fowl)		3.5	18	159	31	1,427	Duke et al. (1973)
Great-horned owl (*Bubo virginianus*)	Calc. value	2.2	—	—	—	1,615	Dicker and Haslam (1972)
Domestic fowl	Calc. value	2.2	—	—	—	3,150	
Ostrich (*Struthio camelus*)	Resp. evaporation	0.6	6.5	100	20	100,000	Crawford and Schmidt-Nielsen (1967)

obvious that most granivorous birds must drink water in increasing amounts as a function of decreasing body weight (see Fig. 2.1.).

A number of observations of rate of evaporation published since the work of Crawford and Lasiewski (1968) are summarised in Table 4.1. Inspection of this table shows good agreement between the actual observations of the recent experiments and predicted values (see Fig. 4.2).

The interesting question is whether the rate of evaporation, when water is freely available and the birds are without thermal stress, deviates in any characteristic manner for any particular group such as xerophilic birds, birds of prey, special families or other groups. No such correlation is, however, obvious. In order to discover whether any adaptation exists in ability to use water for evaporative cooling under heat stress or to conserve water during dehydration the birds must be exposed to these challenges. Sects. 4.2 and 4.3 will deal with these problems.

4.1.3 Other Questions in Relation to Evaporation

Three problems are of interest in relation to evaporation as such. First, the rate of evaporation from the skin in comparison with loss from the respiratory tract; second, the role of counter-current exchange systems for conservation of heat and water in the respiratory tract, and third, the relationship between rate of evaporation and water excretion by cloacal discharge.

An earlier assumption was that cutaneous evaporation played little role in birds because of the absence of sweat glands. This notion can no longer be maintained. In most species for which observations have been made the cutaneous fraction of total evaporation actually surpassed 50% of total rate of evaporation (Table 4.2). Cutaneous evaporation can be measured in small birds by separating the head region from the rest of the body in the flow chamber by means of a rubber collar. The observations on the ostrich (Schmidt-Nielsen et al. 1969) were made on a small fraction of the skin with probably a high relative humidity. The estimate could well be too low.

The experiments of Lee and Schmidt-Nielsen (1971) on the zebra finch were repeated on dehydrated birds. The respiratory evaporation remained identical, but the cutaneous evaporation fell to 27% of total rate of evaporation, a reduction to almost 50% of the value (Table 4.2) observed in watered birds (see further Sect. 10.1). The mechanism for this conservation of water, which may be the same that reduces evaporative water loss in other dehydrated desert birds, remains unknown. It is possible to speculate that feather posture also affects osmoregulation if dehydrated birds sacrifice temperature regulation by decreasing their layer of insulation, just as panting can be suppressed (see p. 156). McFarland and Baher (1968) noticed that Barbary doves (*Streptopelia risoria*), which raise their feathers when cold and sleek them when hot, under dehydration let the feathers stay raised at a higher ambient temperature. This may also decrease evaporation as the water–vapour gradient may become less steep. It should finally be noted that a regulated and quantitatively important cutaneous evaporation is one way, like panting and gular flutter, of dissociating respiratory gas exchange and evaporative cooling.

Table 4.2. Cutaneous water loss as fraction of total rate of evaporation

Species	Total rate of evaporation % of body weight/day	Cutaneous evaporation % of total evaporation	References
Zebra finch (*Poephila guttata*)	16.4	48	Lee and Schmidt-Nielsen (1971)
Zebra finch[a]	9.9	27	
Zebra finch	21.4	63	Bernstein (1971a)
Budgerigar (*Melopsittacus undulatus*)	21.6	59	
Village weaver (*Textor cucullatus*)	16.1	51	
Painted quail (*Excalfactoria chinensis*)	11.3	45	
Painted quail (*Excalfactoria chinensis*)	12.7	59	Bernstein (1971b)
Poorwill (*Phalaenoptilus nuttallii*)	14.2	51	Lasiewski and Bernstein (1971)
Roadrunner (*Geococcyx californicus*)	7.0	51	
Ostrich (*Struthio camelus*)	5.3	2	Schmidt-Nielsen et al. (1969)

[a] Dehydrated birds

If the respiratory gas exchange should proceed with the smallest accompanying water loss, the lowest possible temperature of exhaled air would be an advantage as the air can be no more than fully saturated. The pressure of saturated water vapour and therefore the content of water in the exhaled air is reduced to 50% from body temperature to normal room temperature. That the upper airways have lower temperature than core body temperature when the birds resort to evaporative cooling was found in the domestic fowl (Hutchinson 1955), the pigeon (Calder and Schmidt-Nielsen 1967), and the ostrich (Schmidt-Nielsen et al. 1969). Lasiewski and Snyder (1969) found surface temperature in the upper airpassages to be 0.9°–5.1°C lower than body temperature in nestling double crested and pelagic cormorants (*Phalacrocorax auritus* and *pelagicus*). Schmidt-Nielsen et al. (1970) examined in detail the problem of whether condensation and evaporation of water in the upper airways of birds actually function significantly in conservation of heat and water by counter-current exchange. The object was to observe whether any special adaptation exists in desert birds as compared to non-desert birds. The inhaled air is heated to body temperature and saturated with water vapour in the upper respiratory tract. This makes the temperature of the surfaces fall below body temperature. When exhaled air passes these cooler surfaces the air is cooled and water is condensed thus retaining both heat and water. These investigators examined this system in seven species, including the budgerigar the cactus

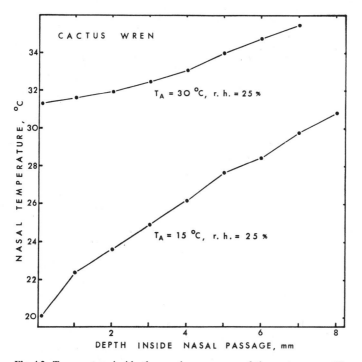

Fig. 4.3. Temperature inside the nasal passageway of the cactus wren. The temperature at various depths inside the nasal passageway of the cactus wren (*Campylorhynchus brunneicapillum*) is shown at 15°C and 30°C ambient temperature. At the lower temperature the gradient was 1.2°C/mm depth. The low temperature leads to a quantitatively important recondensation of water from the exhaled air. Reproduced with permission from Schmidt-Nielsen et al (1970)

wren (*Campylorhynchus brunneicapillum*), the domestic pigeon and domestic duck. The authors found no difference between birds from desert and non-desert habitats and even the pigeon, which is 5 to 10 times larger than the smaller birds, did not deviate from the general pattern. The conservation of water, however, was of importance. In the cactus wren at 30°C ambient temperature and a relative humidity of 25% the temperature declined 0.6°C per mm in the first 7 mm of the nasal passage. Compared to the amount of water that would have been lost if the air were exhaled at body temperature, the recovered amount was 49%. At lower ambient temperature the recovery rate was greater; 74% of water conservation was calculated at 15°C as the temperature decreased by 10°C in the last 8 mm of the nasal passage (Fig. 4.3).

Wunder and Trebella (1976) have examined whether the presence of nasal tufts affects water loss in a similar manner in the common crow (*Corvus brachyrhynchos*). They observed, however, identical rates of evaporation in controls and in birds in which the tufts were cut at both 5°C and 37.5°C ambient temperature.

Simultaneous measurements of rate of evaporation and cloacal excretion of water have been made for a few species. This allows comparison of the relative roles of the two avenues of water loss. In other experiments intake of water and food have

Table 4.3. Evaporation as fraction of total water excretion

Species	Water ad lib. (%)	Dehydration	References
Budgerigar	—	67	Krag and Skadhauge (1972)
(Melopsittacus undulatus)			
Zebra finch	62	61	Skadhauge and Bradshaw (1974)
(Poephila guttata)			
Zebra finch	—	67	Lee and Schmidt-Nielsen (1971)
Stark's lark	—	79	Willoughby (1968)
(Spizocoryx starki)			
Grey-backed finch-lark	—	84	Willoughby (1968)
(Eremopterix verticalis)			
Bobwhite	49	60	McNabb (1969a)
(Colinus virginianus)			
California quail	51	68	McNabb (1969a)
(Lophortyx californicus)			
Gambel's quail	44	64	McNabb (1969a)
(Lophortyx gambelii)			
Great-horned owl	48	—	Duke et al. (1973)
(Bubo virginianus)			
Domestic fowl	51	—	Dicker and Haslam (1972)
Domestic duck	—	72	Stewart (1972)

been measured, production of metabolic water calculated, and cloacal discharge of water measured. This allows a reasonably accurate indirect estimate of the rate of evaporation. The data for several species are recorded in Table 4.3. It will appear that the general pattern is nearly 50% of evaporative loss when water is available ad libitum which is increased, relatively, to 66% or more in dehydrated birds or birds on minimum water, as measured on the small xerophilic seed-eaters. It should be emphasized that the increase in evaporation is relative. Dehydrated birds either do not change or decrease the rate of evaporation (see p. 51), but renal and intestinal output of water is reduced even more. In the species of quail studied by McNabb (1969a; see Table 4.3) increases in relative rate of evaporation originated in a statistically unchanged rate of evaporation, but with a reduction of cloacal discharge of water to less than 50%. In these birds it is justified to consider evaporation as an unavoidable water loss with the entire regulation resting on kidney and cloaca.

4.2 Reaction to Heat Loads

When challenged by overheating, birds have three mechanisms by which rate of evaporation can be increased, and one by which necessity of resort to evaporative cooling is avoided. The rate of evaporation may be increased by panting, by gular flutter or by exposing feet and skin to air. Evaporation may be curtailed by allowing

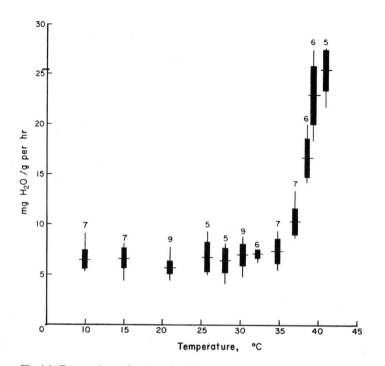

Fig. 4.4. Evaporation as function of ambient temperature. The effect of ambient temperature on the rate of evaporation is shown in the sage sparrow (*Amphispiza belli nevadensis*). The thermoneutral zone is extended until 35°C. The temperature at which rate of evaporation increases from the basal level, and the maximal value, are the bases of the data in Table 4.1. Reproduced with permission from Moldenhauer (1970)

body temperature to increase. This increases heat dissipation by radiation, convection, and conduction. The ability to withstand a heat load can be quantitated, first, as the maximal rate of evaporation related to total heat production and, second, as the maximal ambient temperature at which the rate of evaporation begins to increase. Many species have been subjected to laboratory experiments in which heat stress has been applied by high ambient temperature, usually in a range of 30°–45°C. Rate of evaporation, oxygen consumption, and often body temperature, have been measured. Evaporation in a typical experiment on the sage sparrow is shown in Fig. 4.4. In this, as in most other experiments, the rate of evaporation increased exponentially from an upper critical ambient temperature. The maximal fractional rate of evaporation and fraction of heat dissipated by evaporative cooling of a number of species are recorded in Table 4.1. In the majority of the experiments carried out since Lasiewski et al. (1966a, b) pointed out that the flow rate in the chamber must be high to avoid limitation by high relative humidity, most species have been shown to lose more heat by evaporation than the body produces, i.e. typically about 150% (see Table 4.1). This means that in an atmosphere of low relative humidity birds can tolerate 40°–42°C ambient temperature and still receive an extra heat load predominantly by solar radiation. The role of relative humidity should be stressed. In a near-saturated atmosphere, such as in the tropical rain forest, birds may not be able to tolerate more than about 30°C (King and Farner 1964). Differences in experimental conditions, particularly a relatively high rate of relative humidity, may be the most important reason for differences in estimates of maximal evaporation rates. Heat tolerance is thus not well characterised by this parameter.

Inspection of Table 4.1 shows that maximal rates of evaporation of approximately 60% of body weight per day are general. This is 5 to 15 times the rate at thermoneutrality. Thermal stress thus necessitates access to free water. However, daily drinking rates cannot be expected to reflect these maximal turnover rates since not only the heat stress is absent at night, but the rate of metabolism is reduced too (Aschoff and Pohl 1970). Furthermore, it is noteworthy that evaporative cooling due to high ambient temperature alone (resting birds in the shade) begins at temperatures as high as about 37°C. Since this temperature occurs only in the hottest deserts for more than 4–6 h of the day, there is not always a need for evaporative cooling. Xerophilic seed-eaters, which maintain their rate of evaporation close to the rate of production of metabolic water, will therefore, even when heat-stressed, have only a limited need for extra intake of free water.

4.3 Reaction to Dehydration

For desert mammals, particularly as demonstrated by experiments on the camel (Schmidt-Nielsen et al. 1957), it is known that evaporative cooling is avoided by diurnal increase in body temperature allowing heat loss by radiation at night. It has also been observed, especially on gazelles in East-African deserts, that dehydration suppresses panting with a resulting increase in body temperature (Taylor 1972). The problem is how birds react to simultaneous heat stress and lack of water. The possible ways of reaction are numerous, and as the subject has been little studied, both new mechanisms and quantitative evaluation of the role of known systems are likely to emerge in the future.

Known mechanisms that decrease rate of water loss are: (1) Reduction of the basic rate of evaporation. (2) Avoidance of evaporative cooling either by reduced panting, or gular flutter, or increased body temperature. (3) Reduced diurnal activity. The resulting lower metabolic rate and lower food intake decreases water loss. (4) Torpor or dormancy.

A reduced rate of evaporation has been observed in several xerophilic seed-eaters maintained on a diet of dry seeds such as the budgerigar (Cade and Dybas 1962; Greenwald et al. 1967; Lee and Schmidt-Nielsen 1971), the zebra finch (Cade et al. 1965), Stark's lark (*Spizocoryx starki*), the grey-backed finch-lark (*Eremopterix verticalis*), the silver-bill finch (*Lonchura malabarica*) (Willoughby 1968, 1969), and the species of quail studied by McNabb (1969a).

Both the rate of panting and intervals of gular flutter, which usually occurs at resonance frequency, are functions of body temperature (e.g. Lasiewski and Bartholomew 1966). A reduction of these parameters in dehydrated animals is thus, in principle, easy to quantify. There are, however, only few published measurements. Crawford and Schmidt-Nielsen (1967) observed a decreased rate of respiration and increased cloacal temperature after daily intervals of heat stress during a 7-day period of dehydration in the ostrich. On the seventh day (last day before rehydration) the rate of respiration had decreased to 48% of the level before dehydration, and the cloacal temperature during the period after heat stress was increased by 3.1°C.

In other observations on the ostrich (Louw et al. 1969; Louw 1972), when the birds were kept under desert conditions, panting was shown to be the last resort to avoid overheating. If a wind was blowing the birds did not pant at an ambient temperature of 36°C, but raised their feathers to allow convective cooling (surface temperature, 36.5°C). If there was no wind they panted vigorously; ventilation rate increased by a factor of 10 at 35°C. Skadhauge (1974a) observed that dehydrated emus did not pant at 38°C, but began to do so within 15 min after rehydration.

Reduced diurnal activity has been observed regularly in dehydrated birds. Decreased activity lowers water loss both by evaporation and by cloacal discharge. McFarland and Wright (1969) found an almost linear relationship between cloacal water loss and food intake in dehydrated Barbary doves. Also in the small desert species, budgerigar and zebra finch, food intake is lower during dehydration, and activity is decreased (Skadhauge and Bradshaw 1974; Krag and Skadhauge 1972; Greenwald et al. 1967).

Concerning torpor MacMillen and Trost (1967) observed pronounced nocturnal hypothermia in the Inca dove (*Scardafella inca*) deprived of food or water, or both. The nocturnal evaporation rate in dehydrated birds was reduced to one-third of the diurnal value.

The question of how dehydration hyperthermia is regulated in birds should be elucidated. Body temperature is, as in mammals, regulated from the hypothalamus (e.g. Simon-Oppenman et al. 1978), but the mechanism that controls hyperthermia is unknown.

Chapter 5

Function of the Kidney

In all terrestrial vertebrates the kidney excretes water, salts, nitrogenous waste, and hydrogen ions; and it also produces hormones. This chapter is divided in three main sections on excretion of salt and water, action of antidiuretic hormones, and nitrogen excretion.

5.1 Excretion of Water and Salts

The description of the filtration-resorption kidney of birds begins with a survey of anatomy and circulation, and measurement of renal plasma flow. Regulation of the rate of glomerular filtration (GFR) is discussed, followed by description of excretion of salt, water, and urea as fractions of the filtered rate in different osmotic situations. Maximal concentrating ability is outlined, and related to renal morphology. This involves a description of the range of solute concentrations of the ureteral urine that are regurgitated into the cloaca.

Precise measurements of GFR and ureteral excretion of NaCl and water have been carried out in several species on ad-libitum intake of water, during forced hydration, during dehydration, and after salt loading. The following species have been investigated: The domestic fowl (Dantzler 1966; Skadhauge and Schmidt-Nielsen 1967a), a desert quail (*Lophortyx gambelii*) (Braun and Dantzler 1972, 1975; Braun 1976), the domestic turkey (Vogel et al. 1965), the European starling (Clark et al. 1976; Braun 1978); the budgerigar (Krag and Skadhauge 1972), and the domestic duck (Holmes et al. 1968). The main osmoregulatory problems would seem to be: How stable is GFR in birds as a function of the level of hydration and salt intake? How large is the range of tubular resorption of water and salt? What is the maximal concentrating ability? How do the "reptilian" type and "mammalian" type nephrons function during various types of osmotic stress? What is the role of the supply of renal portal blood to the peritubular capillaries? What is the effect of uric acid as the main end-product of nitrogen metabolism?

5.1.1 Renal Anatomy

The renal functions, particularly renal plasma flow (RPF), GFR, and concentrating ability (urine to plasma osmotic ratio), are in all vertebrate groups closely related to the anatomy of the kidney. This applies to nephrons as well as to blood vessels. The subject of renal morphology and circulation will therefore be reviewed with special

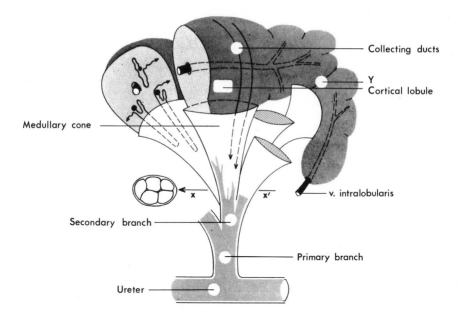

Fig. 5.1 The organisation of the avian kidney. The cortical lobules contain reptilian-type nephrons (without loops of Henle), whereas the loops of Henle of the mammalian-type nephrons extend into the medullary cones. *Y* illustrates a region in which two cortical lobules supply the same adjacent medullary cone. A transverse section at the plane x-x' shows how the medullary cones converge on the secondary ureteral branch. Modified and reproduced with permission from Johnson (1974)

attention to the mechanism of formation of concentrated urine. This requires hairpin loops, both of the nephrons (loops of Henle) and the blood vessels (vasa recta). In a later section (see p. 73) an attempt to correlate structure and function will be made. The functional role of the renal portal circulation will be discussed (see p. 57).

The avian kidney is situated dorsally in the abdomen suspended like a hammock in strings of connective tissue. These structures are so closely interwoven that it is usually impossible to remove the fresh organ in toto. Each kidney is divided into three lobes. The ureter runs centrally along the ventral side. On the surface of the kidney several lobules are visible. They are connected to the ureteral branches by pyramids of collecting ducts. These ducts merge in the medullary cones close to the central ureteral branch (Fig. 5.1). Each lobule may send loops of Henle and collecting ducts to several medullary cones (Johnson et al. 1972). On the outside of the lobules lie the veins that carry portal blood to the renal tissue, i.e. vv. interlobulares. The arterial blood supply enters the lobule from the centre (Fig. 5.3) and here, also, is located v. centralis or v. intralobularis. This vessel carries blood from the kidney. The lobules are thus surrounded by a capillary network that receives a mixture of arterial (post-glomerular) and portal blood.

In lower vertebrates, which have low rates of filtration (Schmidt-Nielsen 1964; Schmidt-Nielsen and Skadhauge 1967), the portal circulation assumes a greater role, culminating in the aglomerular teleosts in which there is no glomerular

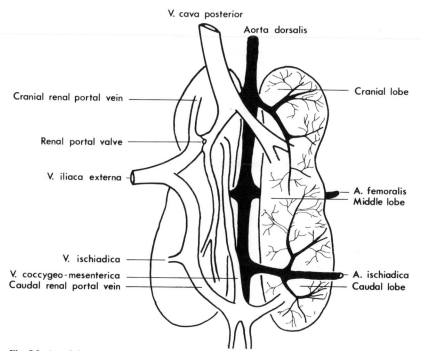

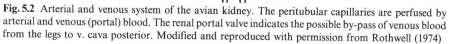

Fig. 5.2 Arterial and venous system of the avian kidney. The peritubular capillaries are perfused by arterial and venous (portal) blood. The renal portal valve indicates the possible by-pass of venous blood from the legs to v. cava posterior. Modified and reproduced with permission from Rothwell (1974)

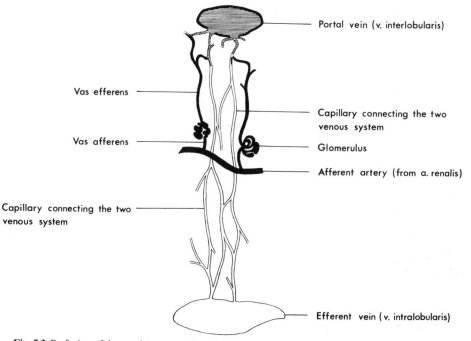

Fig. 5.3. Perfusion of the renal cortex. Both arteries and portal veins deliver blood to the cortical tissue. The cortical nephrons (not shown) in the domestic fowl are perfused by near equal amounts of arterial and venous blood (see p. 59)

filtration. The absence of a portal circulation in mammals seems to be an important feature. In birds a portal flow across the medullary tissue would presumably counteract the build-up of papillary hypertonicity. This depends, as stated above, on the presence of a counter-current system with a double set of hairpin loops, the loops of Henle and the vasa recta. For these reasons birds present a problem. They can concentrate urine above the osmolality of plasma, yet they have a large portal circulation.

When the morphology of the avian kidney is viewed in the light of these facts, birds seem to have placed a mammalian medullary cone upon a reptilian lobule. In the lobule the nephrons run towards the central lobular vein, turn around towards the periphery of the lobule again, and join the collecting ducts (see Fig. 5.1). This arrangement is typical for the reptilian kidney (Spanner 1925). It would seem impossible to build up a hyperosmotic area in these parts of the kidney, since blood flows all along the nephron towards the central vein. In the medullary cone areas, however, loops of Henle appear (see Fig. 5.1) (Berger 1966; Braun and Dantzler 1972; Feldotto 1929; Huber 1917; Johnson and Mugaas 1970a, b; Siller and Hindle 1969; Sperber 1960; Tchang 1923), and a counter-current system may be expected to operate. Precise information on the blood supply to this region is lacking, but it seems to be exclusively postglomerular (systemic) blood (Sperber 1949) i.e. non-portal. From the arterial injection studies of Siller and Hindle (1969) on the domestic fowl it appears that the vasa recta system is well developed in the medullary cones. Evidence of presence of hypertonicity in this region and thus, by inference of the operation of a counter-current system has been found (see p. 73). The collecting ducts have, compared with mammalian nephrons, a fairly large diameter, 30–60 μm, and are located in the medullary cone surrounded by the loops of Henle (Braun and Dantzler 1972; Johnson and Mugaas 1970a; Tchang 1923). According to the neoprene casts of Siller and Hindle (1969), the collecting ducts increase in diameter to approximately 100 μm, as they merge towards the medullary cone.

The fresh kidney is very fragile in birds, and the medullary cones cannot easily be distinguished on the cut surface or in the fixed kidney. Consequently, the study of the possible correlation between the anatomy of the medullary cone and concentrating ability is not as easily carried out in birds as in mammals. Some information is available, particularly as a result of the meticulous work of Johnson and Mugaas (1970a, b). They observed that the number of medullary lobules per milligramme of kidney cortex is larger in effective water conservers, and that there is some indication of large medullary cones being associated with better concentrating ability. Poulson (1965) also observed a better correlation between concentrating ability and number of loops of Henle per unit area of the medulla rather than papillary length in house finches and savannah sparrows. These findings will be discussed later (see p. 74).

5.1.2 Circulation of the Kidney

A common renal artery does not exist in birds, but arterial blood coming from the aorta enters the kidney by a. renalis anterior, and by branches from a. ischiadica and a. femoralis (see Fig. 5.2). The intrarenal arteries follow the efferent veins closely. The glomeruli are smaller than in mammals—about 60 μm in diameter—but more numeros. They were considered as "poorly developed" by Marshall and Smith (1930). However, as pointed out by Sperber (1960), such judgement should be passed only on functional evidence. The glomerular filtration rate, at least in the domestic fowl and domestic duck, is not lower than in mammals of comparable size. The flow in the single nephron GFR in Gambel's quail is 15.8 nl/min in mammalian type nephrons with long loops, and 6.4 nl/min in reptilian type nephrons (Braun and Dantzler 1972). The former value is only slightly smaller than in mammals. In the rat values of 20–40 nl/min are generally found. Thus, there would seem to be no compelling reason for the claim that the glomeruli are poorly developed in birds.

Portal blood comes from the leg. The v. iliaca externa crosses the ventral surface of the kidney embedded in the renal tissue (see Fig. 5.2). It divides into two large branches to the kidney and forms an anastomosis with the efferent renal vein leading to the v. cava posterior. The anastomosis has a valve partly composed of smooth muscle. This suggest that the amount of blood bypassing the kidney may be regulated. The branches to the kidney terminate as the vv. interlobulares supplying the peritubular capillaries (see Fig. 5.3).

That the portal blood in fact supplies the kidney tubules was convincingly demonstrated by Sperber (1948, 1949) who studied the circulatory system of the kidney of several birds and performed physiological experiments on the domestic fowl. He injected various organic coloured substances such as phenol red, which are subject to tubular secretion, into the leg vein, and observed that dye appears earlier and in higher concentration on the side of the injection. Later, careful radiographic studies by Akester (1964, 1967), also on the domestic fowl, demonstrated that portal blood flows into the entire kidney. Akester (1967) also showed that the portal blood may follow three shunts to the systemic circulation bypassing the kidney: (1) via the renal portal valve and the v. cava; (2) via the caudal portal vein and the v. coccygeo-mesenterica (see Fig. 5.2), which sends blood to the liver; and (3) via the cranial renal portal vein and the vertebral venous sinuses. Portal blood may even reach the contralateral kidney before entering the systemic circulation. Later studies have confirmed that blood is shunted from the femoral and the renal portal vein to the liver (Sturkie and Abati 1975; Sturkie et al. 1977, 1978).

Do the anastomoses between portal and systemic circulation have any functional significance? This question has received considerable attention since the discovery of the valve between the renal portal vein and the vein (v. renalis revehens or common iliac vein) leading to v. cava posterior (Spanner 1939). Sperber (1960) thought that the blood flow to the kidney would bypass the kidney through the shunt when the bird was performing muscular work, thereby increasing blood flow through the leg to a higher level than the kidneys demanded. This regulatory system would thus be expected to operate maximally in a running bird such as the domestic fowl. There have been no functional studies of this problem, but Rennick and Gandia (1954) demonstrated that the smooth-muscle fibres in the valve reacted to acetylcholine and to histamine with contraction, and to epinephrine with relaxation. A rich supply of autonomic nerves has been observed in the valve of the domestic fowl (Gilbert 1961), and shown by histochemical techniques to contain adrenergic and cholinergic fibres (Akester and Mann 1969). It should be noted that another running bird, the ostrich, has a well-developed anastomosis with three valves (Oelofsen 1977). The nandu (*Rhea americana*) has, however, only one (Sperber 1948).

It is possible that the state of hydration is a factor determining how much blood from the leg will bypass the kidney. Skadhauge (1964) found that PAH (para-aminohippurate) infused into the leg vein on one side in hydrated hens at a rate sufficiently low to be cleared by the kidney in one passage, according to the estimates given by Sperber (1960) and by Dantzler (1966), was excreted with only two parts from the ipsilateral ureter for one part from the contralateral ureter. Other investigators using non-hydrated hens found a 7–10 fold higher excretion rate on the ipsilateral than on the contralateral side (Rennick et al. 1956; Volle et al. 1962). The low ipsilateral excess in the experiments of Skadhauge (1964) may be due therefore to a larger fraction of the portal blood going directly to the systemic circulation.

5.1.3 Renal Plasma Flow

Renal plasma flow (RPF) is generally estimated by the clearance of a substance that is subject to tubular secretion, such as PAH or phenol red. Clearance should be so effective that the substance entering the kidney leaves it in the urine within the same unit time. If this ideal situation does not obtain, the error should be small and constant. In mammals a correction factor of 0.9 of the RPF estimated by direct calculation of the PAH clearance will account for the error. The error is presumably equally small in birds, but no results from direct control experiments are available.

Clearance measurements for determination of RPF and GFR are only reliable if the plasma concentration and the urine flow rate remain relatively constant during the experimental period. If this is not so, for example during the waxing or waning of antidiuresis, the apparent excretion of the substance in the urine will be retarded or accelerated as compared to the normal delay during passage through the ureter, and errors will appear in the estimate of the clearance. If the object is to calculate RFP or GFR during action of antidiuretic hormones, an integration may be carried out by taking the average clearance for the experimental periods during the total antidiuresis, as judged for example by deviation of urine osmolality from the control value. This method was used by Skadhauge (1964). Ames et al. (1971) have applied a correction for ureteral delay during the development of antidiuresis (see p. 80) for determination of GFR.

Renal plasma flow, as determined by PAH clearance, has been measured in domestic birds (Table 5.1) with rates of about 40 ml/kg · min found in the domestic fowl. Very few measurements in other species have been published. It would be interesting to compare the RPF/metabolism ratio for herbivorous and for carnivorous birds, since the latter, ingesting a high protein diet, would presumably produce more uric acid for excretion. No change in RPF was detected during the action of small doses of antidiuretic hormone (Skadhauge 1964).

Sapirstein and Hartman (1959) have attempted to estimate the cardiac output and renal blood flow by another method, i.e. by rapid injection of RB[86] in the neck vein and counting of the activity in multiple samples of carotid blood within the next few seconds, and activity of the organs when the birds were killed 2 min after

Table 5.1. Renal plasma flow (RPF). PAH clearance, unless otherwise stated

Species	Notes	RPF ml/kg· min Average values	References
Domestic fowl	Phenol red clearance	25	Pitts (1938)
Domestic fowl	Uric acid clearance	27	Shannon (1938b)
Domestic fowl		40	Nechay and Nechay (1959)
Domestic fowl		50	Orloff and Davidson (1959)
Domestic fowl		29	Sperber (1960)
Domestic fowl	„Diodone" clearance	18	Sykes (1960)
Domestic fowl		40	Skadhauge (1964)
Domestic fowl	Rb[86] uptake[a]	23	Sapirstein and Hartman (1959)
Domestic fowl		37	Osbaldiston (1969b)
Domestic fowl	Laying hen, labelled microspheres[a]	18	Boelkins et al. (1973)
Domestic fowl	Laying hen, labelled microspheres[a]	31	Wolfenson et al. (1974)
Domestic duck	Fresh water	21	Holmes et al. (1968)
	Hypertonic saline	22	
Domestic duck	Fresh water	16	Stewart et al. (1969)
	Hypertonic saline	15	

[a] Does not include portal flow

the injection. The average renal blood flow was determined as 15.2% of the cardiac output. Arterial renal blood flow was calculated with 33 ml/kg·min. If the haematocrit is 30% in the domestic fowl (Bond and Gilbert 1958), the arterial RPF is $(70/100 \times 33 = 23$ ml/kg·min, which is approximately half the total renal plasma flow as estimated by the PAH clearance method. Two groups have used radioactively labelled microspheres in a modification of the technique of Sapirstein and Hartman. Boelkins et al. (1973) in investigations with white leghorn pullets found that the fraction of cardiac output of the kidneys is 10.0%, which is a renal arterial blood flow of 17.6 ml/kg·min. In the domestic goose Hanwell et al. (1971a) observed 15% of cardiac output going to the kidneys. In laying hens Wolfenson et al. (1978) measured the arterial blood flow to the left kidney to be 2.2 ml/g·min and to the right kidney 1.9 ml/gm·min. As the average body weight of the birds was 2.0 kg and the weight of the kidney 14.9 g on average, a renal blood flow of 30.9 ml/kg·min was calculated.

Odlind (1978) studied the distribution of blood flow in the renal portal system of laying hens and observed that an average of 44% of the portal blood from the v. iliaca externa flows to the ipsilateral kidney. The flow pattern was highly variable in spite of constant experimental conditions. Local vasoconstriction occurred frequently in the portal vessels in the kidney.

5.1.4 Glomerular Filtration Rate (GFR)

Several investigators have measured GFR in the domestic fowl, but only few measurements have been made in other species, including the domestic turkey,

Gambel's quail, European starling, budgerigar, the domestic duck, and, among marine birds, and the glaucous-winged gull. In most experiments the GFR was determined as clearance of inulin. The ideal marker of clomerular filtration should be filtered with the same concentration in the filtrate as in the water phase of plasma, and be neither secreted nor subjected to resorption or back diffusion or metabolism in the tubules. If these conditions are fulfilled and the plasma level and urine flow is held constant throughout the experimental period, GFR may be expressed by the clearance of the substance. A clearance independent of plasma concentration may be considered a good proof of the validity of the assumptions. This test has been performed on inulin in the domestic fowl (Pitts 1938; Shannon 1938a; Lambert 1945). It is further confirmed by the finding of equal inulin and polyethylene glycol, 4000 M (PEG) clearances in the domestic fowl (Hydén and Knutson 1959).

The measured values of GFR are recorded in Table 5.2. It will appear that for all avian species investigated most values fall within a range of 1–3 ml/kg·min in birds not subjected to osmotic stress. For the domestic fowl the GFR is almost of the same magnitude as in mammals of similar weight, e.g. rabbit and cat. If the portal blood flow, as approximately calculated earlier (see p. 59), is assumed to be 50% of total blood flow to the kidney, the filtration fraction can be calculated at around 15%. This is slightly less than the 19%–20% recorded for man and rat. The GFR values measured in early investigations (Shannon 1938a; Pitts 1938; Pitts and Korr 1938; Korr 1939) appear somewhat lower than those from more recent experiments. This is probably due to the less stressful technique employed. Smaller doses of anaesthetics, maintenance of body temperature, blood taken from indwelling catheters rather than by heart puncture are factors that may serve to explain the differences.

When the measured GFR of the different species are compared no consistent picture emerges. It is noteworthy that birds with salt glands, e.g. domestic duck and the glaucous-winged gull, fall within the range of the galliform birds. The budgerigar has a relatively high rate of filtration whereas fairly low values have been observed in Gambel's quail. Desert habitation is, thus, not necessarily associated with a low GFR.

The GFR has been measured as a function of two types of osmotic stress, hydration and dehydration, and salt loading. In the classical studies on the domestic fowl of Pitts (1938) and Shannon (1938a) the GFR was consistently found higher at the highest rates of urine flow. Later studies on domestic fowl (Skadhauge and Schmidt-Nielsen 1967a) and budgerigar (Krag and Skadhauge 1972) demonstrated a significant augmentation of GFR from dehydration to hydration. This is clearly related to the action of argenine vasotocin (AVT) (see p. 81). In the domestic duck Holmes et al. (1968) observed a direct proportionality between urine flow rate and GFR when the birds were on ad libitum intake of fresh water or a hyperosmotic saline solution. This correlation depends, however, almost entirely on a single bird. In principle, correlation of two variables in which one, urine flow rate, is a factor in the other, GFR, is less satisfactory. Correlation of urine flow rate with urine/plasma ratio of the filtration marker is preferable.

The data of Table 5.2 show a pronounced decrease in GFR following an intravenous salt load in domestic fowl (Dantzler 1966; Skadhauge and Schmidt-Nielsen 1967a), domestic duck (Holmes et al. 1968), and Gambel's quail (Braun

Table 5.2. Glomerular filtration rate (GFR). Inulin clearance, unless otherwise stated; PEG = polyethylene glycol

Species	Notes	GFR (ml/kg · min) (mean)	References
Domestic fowl	Hydration	1.84	Pitts (1938)
	High rates of urine flow	2.5	
Domestic fowl	Hydration	2.04	Pitts and Korr (1938)
Domestic fowl	Hydration	1.87	Shannon (1938a)
Domestic fowl	Dehydration	0.6	Korr (1939)
	Normal hydration	1.22	
	Water loading	2.19	
Domestic fowl		1.71	Lambert (1945)
Domestic fowl		2.92	Hydén and Knutson (1959)
	PEG clearance	2.84	
Domestic fowl		2.4	Nechay and Nechay (1959)
Domestic fowl		1.8	Berger et al. (1960)
Domestic fowl		3.0	Sperber (1960)
Domestic fowl		1.84	Sykes (1960)
Domestic fowl	Hydration	2.86	Skadhauge (1964)
Domestic fowl	PEG clearance	2.21	Sanner (1965)
Domestic fowl	Before salt loading	1.23	
	During salt loading	0.5	Dantzler (1966)
Domestic fowl	Hydration	2.12	Skadhauge and Schmidt-Nielsen (1967a)
	Dehydration	1.73	
	Salt loading	2.06	
Domestic fowl	Antidiuresis	0.72	Osbaldiston (1969a)
Domestic fowl, 9 days old	EDTA Cr^{51} clearance	1.76	Cooke and Young (1970)
Domestic fowl	Peak AVT antidiuresis	2.71	Ames et al. (1971)
Domestic fowl		3.03	Svendsen and Skadhauge (1976)
Domestic duck	Fresh water	2.5	Holmes et al. (1968)
	Hypertonic saline	2.1	
Domestic duck	Fresh water	2.2	Stewart et al. (1969)
	Hypertonic saline	2.1	
Domestic turkey	Creatinine clearance	1.32	
	Hydration	2.6	Vogel et al. (1965)
Glaucous-winged gull	Before salt loading	2.3	Hughes quoted from
(*Larus glaucescens*)	During salt loading	0.74	Schmidt-Nielsen (1964)
Gambel's quail	Mannitol diuresis	0.88	Braun and Dantzler (1972)
(*Lophortyx gambelii*)	Mannitol diuresis	1.11	Braun and Dantzler (1974)
	Hydration (hypo-osmotic glucose-saline)	1.39	Braun and Dantzler (1975)
	Mannitol diuresis	1.23	Braun (1976)
	Mannitol + salt loading	0.45	Braun (1976)
Budgerigar	Hydration	4.43	
(*Melopsittacus undulatus*)	Dehydration	3.24	Krag and Skadhauge (1972)
European starling	Mannitol diuresis	7.40	Clark et al. (1976)
(*Sturnus vulgaris*)	Mannitol diuresis	2.92	Braun (1978)

1976). The decrease in GFR after a salt load is dependent on the magnitude of the load. In general, the salt load induced an osmotic diuresis as observed in domestic fowl (Dantzler 1966; Skadhauge and Schmidt-Nielsen 1967a), domestic pigeon (Scothorne 1959), Pekin duck (Holmes et al. 1961b; Scothorne 1959), and double-crested cormorant (Schmidt-Nielsen et al. 1958). In the domestic fowl a moderate salt load of 15 mM/kg, which only increased the fraction of filtered NaCl excreted to 7%–8%, did not measurably reduce GFR (Skadhauge and Schmidt-Nielsen 1967a), whereas very high salt loads (20–40 mM/kg), which resulted in an excretion of 30% of filtered Na, led to a reduction of GFR of up to 40% (Dantzler 1966). The mechanism of reduction of GFR seems to be release of arginine vasotocin (AVT) in response to high plasma osmolality (see p. 83). In Gambel's quail salt load reduces the filtration of reptilian type nephrons leading to complete shutdown of individual glomeruli (see p. 65). In the domestic fowl the reduction of GFR after salt loading measured together with maximal tubular resorption of PAH (T_{max}PAH; Dantzler 1966), demonstrated a proportional reduction. This was interpreted to indicate that some glomeruli stopped filtering completely. If all glomeruli filtered less and the functional tubular capacity was unchanged (compare p. 64), T_{max}PAH would have remained constant.

5.1.5 Renal Excretion of Sodium and Water

The simultaneous determinations of urine flow rate, and the concentrations of Na and inulin, or another marker, in urine and plasma allow calculation of the glomerular filtration rate, the fractional resorption of water, and the fraction of filtered Na that is excreted. In Table 5.3 the available evidence is assembled according to three functional states. First, control studies with the birds in a mild state of hydration after an oral water load or an infusion of either mannitol or a hypo-osmotic glucose-NaCl solution; second, dehydration; third, hyperosmotic salt loading. It presents a rather uniform picture of glomerular filtration rate and tubular resorption of water. From hydration to dehydration GFR is reduced only moderately, by 23% in the domestic fowl (Skadhauge and Schmidt-Nielsen 1967a) and by 37% in the budgerigar (Krag and Skadhauge 1972). Concomitantly, tubular resorption of water increases from approximately 80% in the overhydrated state, to 95% during mild hydration, and to more than 99% in the dehydrated state. All the species investigated show this wide range of excretion and resorption of water.

Sodium excretion has been found generally to be about 1% of the filtered load unless an external NaCl load is given, which is similar to the condition in mammals. The mechanisms that cause changes in GFR and urine flow rate are discussed in Chapter 5.2, and in the discussion of single-nephron GFR (see p. 65). A higher fractional excretion rate of Na has been observed during mannitol diuresis (Dantzler 1966; Braun and Dantzler 1972). This is to be expected since mannitol causes osmotic diuresis. During salt loading, particularly with loads higher than 20 mM/kg body weight, two reactions were observed. First, the GFR is reduced and second, the amount of excreted Na is increased to approximately 20% of the filtered load. Concomitantly with the decrease in GFR, T_{max} for glucose and for PAH are reduced (Dantzler 1966; Braun and Dantzler 1972). This indicates that the

Table 5.3. Fractional excretion of filtered sodium and water, chloride and potassium

Species	Notes	Urine osmolality mOs	Urine flow rate ml/kg h	GFR ml kg min	% H₂O absorbed	% Na excreted	% Cl excreted	% K excreted	References
Domestic fowl	Dehydration	538	1.08	1.73	99.0	1.52	0.84	16.2	Skadhauge and Schmidt-Nielsen (1967a)
Budgerigar (Melopsittacus undulatus)	Dehydration	848	1.68	3.24	99.1	0.35	0.32	—	Krag and Skadhauge (1972)
Domestic fowl	Salt loading	362	10.9	2.07	87.2	7.03	8.09	36.5	Skadhauge and Schmidt-Nielsen (1967a)
European starling (Sturnus vulgaris)	—	—	30 (240% of control)	1.1 (100% of control)	80	20	—	—	Dantzler (1966)
	—	785 (peak)				29.3[a]	—	25	Braun (1978)
Domestic duck	—	444	2.61	2.0	96.2	0.2	1.1	34.2	Holmes et al. (1968)
Domestic fowl	Hydration	115	17.9	2.12	85.8	2.40	2.36	13.3	Skadhauge and Schmidt-Nielsen (1967a)
Budgerigar	Control	236	6.30	4.43	97.6	0.20	0.28	—	Krag and Skadhauge (1972)
Domestic turkey	Control	305	3.30	1.32	89.3	4.4	3.2	—	Vogel et al. (1965)
Gambel's quail (Lophortyx gambelii)	2.5% mannitol infusion	374	11.2	0.88	78.2	6.0	—	20	Braun and Dantzler (1972)
	Hydration	—	6.24	1.39	93	<1	—	60	Braun and Dantzler (1975)
European starling	2.5% mannitol infusion	—	19.4	7.40	95.6	—	—	24	Clark et al. (1976)
	2.5% mannitol infusion	549	10.9	2.82	93.0	37.3[a]	—	21	Braun (1978)
Domestic duck	Hydration	241	2.40	2.51	97.3	1.6	0.1	46.6	Holmes et al. (1968)
Domestic turkey	Hydration	136	30.5	2.6	77.7	6.3	7.8	38.1	Vogel et al. (1965)

[a] Determined after uricase treatment

63

GFR is reduced due to complete shutdown of a number of nephrons but not because each nephron filters less than before. If such were the case the tubular capacity for excretion and resorption, as measured by T_{max}, would not have been reduced.

With respect to fractional salt excretion two exceptions from the general pattern (see Table 5.3) are apparent. First, in the only species with a functional salt gland studied in detail, the domestic duck, salt loading by intake of a sea-waterlike solution as the only drinking fluid did not result in a high fractional Na excretion by the kidney. This is because 88% of the load was excreted through the salt gland (Holmes et al. 1968). Second, in the European starling (Braun 1978), about one-third of filtered Na is contained in the uric acid precipitate and only measured after release by treatment with uricase. Braun (1978) realised that such a great amount of Na cannot be lost from the body and suggest that this Na is resorbed in the cloaca.

Finally it should be mentioned that the maximal range of fractional water excretion is greater than the mean values that Table 5.3 suggests. In the domestic fowl (Skadhauge and Schmidt-Nielsen 1967b), the urine to plasma ratio for inulin ranges from 3 to more than 100 from maximal hydration to pronounced dehydration. This implies that the amount of filtered water excreted plunges from more than 30% to less than 1%. Thus, the ability to excrete and conserve water has a wider range than in mammals, and the conservation of water is achieved with a comparatively lower concentrating ability.

5.1.6 Determinations of Single-Nephron Glomerular Filtration Rates (SNGFR)

Knowledge of regulation of GFR in birds has been augmented greatly by the studies of Braun and Dantzler (1972, 1975) and Braun (1976, 1978) who assigned functional roles to the "reptilian type" nephrons without loops of Henle, and the "mammalian type" nephrons with loops of Henle. They were then able to measure the filtration rate of individual nephrons (SNGFR) of both types by the so-called Hanssen technique (Braun and Dantzler 1972), which involves measurement of filtration rate with a non-absorbable marker, Na-ferrocyanide, labelled with an isotope, usually C^{14}. At a given time an arterial injection of a higher dose of unlabelled filtration marker is given as a bolus and after a few seconds, when the bolus is in the middle of the proximal convoluted tubule, the blood supply is stopped by rapid freezing of the kidney, usually by liquid nitrogen. The tissue is thawed gradually, partly digested with HCl, and individual nephrons are isolated under the stereomicroscope. The amount of radioactive substance is removed from the glomerulus to the front of the bolus. (The quantitative collection is possible because Na-ferrocyanide in the frozen state is converted to insoluble Prussian Blue.) The radioactivity collected from the nephron represents the filtered amount for the time from bolus injection to freezing. SNGFR is calculated by division of this amount by the concentration of radioactivity of plasma.

The SNGFR values determined in Gambel's quail (Braun and Dantzler 1972) and in the European starling (Braun 1978) are presented in Table 5.4. In both species the mammalian-type nephrons filter approximately twice as much as the reptilian-type nephrons. Both types of nephrons almost double their filtration rate when the birds are exposed to a water load. But with salt load their reactions differed greatly. The mammalian-type nephrons filter as before but the reptilian-type cease filtration. Concerning the mechanism for this response it is relevant that injections of AVT (see p. 81) mimic the effects of salt loading. As the salt loading increases the plasma osmolality considerably, a part of the response is undoubtedly due to AVT. Furthermore, Braun (1976) addressed himself to the problem how the renal blood flow was regulated to permit the reptilian-type glomeruli to function intermittently. He injected "Microfil", a silicone elastomere, suited for visualisation of vascular networks. Microfil was injected in birds during mannitol infusion and after salt loading. In the control group that received only mannitol the cortical tissue showed uniform filling of the afferent arterioles with the Microfil penetrating through the glomeruli into the peritubular capillaries. After salt loading, however, there was incomplete filling of the fine vasculature in the superficial region of the kidney where the reptilian-type nephrons are located. Few, if any, glomeruli of afferent arterioles were filled. Braun concluded that vasoconstriction at the level of the afferent arterioles reduces GFR.

Functionally this response is most important. Not only is water conserved directly by reduction in GFR, but when a greater fraction of the nephrons that continue to filter have loops of Henle, the concentrating ability should be augmented (see p. 74). To what extent this phenomenon actually determines the maximal renal concentrating ability as a part of a normal physiological mechanism, remains to be investigated.

Summary: From the investigations described above, the avian kidney emerges as a versatile osmoregulatory organ. It allows a high fractional excretion of water (33%) during hydration, and a low fractional excretion (less than 1%) during dehydration. The water conservation is aided by the shutdown of the reptilian-type nephrons. The reduction of GFR is of importance when renal water conservation is continued during the storage of ureteral urine in the cloaca (see Chap. 8). The excretion of "strong" electrolytes may be regulated independently of excretion of water, from 0.1% to about 10% of the filtered amount.

5.1.7 Renal Concentrating Ability

Renal concentrating ability, measured as the maximal osmotic urine-to-plasma ratio in the dehydrated or salt-loaded state, is an important indicator of renal efficacy. Of quantitative importance is also the amount of solute that can be excreted at this concentration. In mammals, the former parameter can be related to relative length of the renal papilla (Schmidt-Nielsen and O'Dell 1961), the latter, in closely related species, to papillary width (Schmidt-Nielsen 1964). In order to correlate structure and function in birds (see p. 73) these parameters must first be determined in a number of species. A fairly large number of observations of the maximal osmotic urine-to-plasma ratio exists (Table 5.5), whereas the amount of solute that can be excreted without reducing urine osmolality has not been precisely determined in any bird. In this section the maximal osmotic urine-to-plasma ratio will be reviewed.

Table 5.4. Single-nephron glomerular filtration rate (SNGFR)

Species	Notes	SNGFR (ml/min)		Fraction of RT open (%)	References
		MT nephrons	RT nephrons		
Gambel's quail (*Lophortyx gambelii*)	2.5% Mannitol diuresis	14.6	6.4	71	Braun and Dantzler (1972)
	Salt load	12.7	0	16	
Gambel's quail	Water load	33.2	11.4	100	Braun and Dantzler (1974)
	AVT (10 ng/kg)	11.3	4.7	52	
European starling (*Sturnus vulgaris*)	2.5% Mannitol diuresis	15.6	7.0	—	Braun (1978)
	Salt load	24.3	10.3	—	

MT = mammalian-type nephrons; RT = reptilian-type nephrons

Table 5.5. Renal concentrating ability

Species	Notes	Osmotic urine-to-plasma ratio	Osmotic and ionic concentration		References
			mOs	mequiv/l	
Dehydration					
Domestic fowl	Average value	1.6	538	Na: 134	Skadhauge and Schmidt-Nielsen (1967a)
Domestic turkey			517	Cl: 108	Vogel et al. (1965)
			492	K: 105	Skadhauge and Schmidt-Nielsen (1967a)
Budgerigar (*Melopsittacus undulatus*)		2.3	848	K: 73	Krag and Skadhauge (1972)
White pelican (*Pelecanus erythrorhynchos*)			580	K: 114	Calder and Bentley (1967)
Roadrunner (*Geococcyx californicus*)			593	K: 75	Calder and Bentley (1967)
Zebra finch (*Poephila guttata*)	Australian specimens	2.8	1,005	K: 135	Skadhauge (1974a)
Galah (*Cacatua roseicapilla*)	10%–15% weight loss	2.5	982	K: 125	

Table 5.5 (continued)

Species	Notes	Osmotic urine-to-plasma ratio	Osmotic and ionic concentration		References
			mOs	mequiv/l	
Singing honeyeater (*Meliphaga virescens*)	—	2.4	925	K: 114	
Red wattlebird (*Anthochaera carunculata*)	—	2.4	917	K: 201	
Senegal dove (*Streptopelia senegalensis*)	—	1.7	661	K: 121	
Crested pigeon (*Ocyphaps lophotes*)	—	1.8	655	K: 162	
Emu (*Dromaius novae-hollandiae*)	—	1.4	459	K: 120	
Kookaburra (*Dacelo gigas*)	—	2.7	944	K: 93	
Ostrich (*Struthio camelus*)	—	2.7	800	K: 139	Louw et al. (1969)
Salt loading					
Birds without salt glands:					
Savannah sparrow (*Passerculus sandwichensis beldingi*)	Drinking saline	5.8	2,000	Cl: 960	Poulson and Bartholomew (1962a)
Savannah sparrow (*Passerculus sandwichensis brooksi*)	Drinking saline	3.2	1,000	Cl: 527	Poulson and Bartholomew (1962a)
House finch (*Carpodacus mexicanus*)	Drinking saline	2.4	850	Cl: 370	Poulson and Bartholomew (1962b)
Domestic pigeon	Cloacal samples			Cl: 246	Scothorne (1969)
Domestic zebra finch			704		E. Skadhauge (unpublished experiments)
Zebra finch	Cloacal samples	2.8	1,027	Cl: 161	Skadhauge and Bradshaw (1974)
Zebra finch	Cloacal samples	1.4 (total electrolytes)		Cl: 224	Lee and Schmidt-Nielsen (1971)
Domestic chicken	Average value (NaCl load: 15 mM/kg)	1.1	362	Na: 164	Skadhauge and Schmidt-Nielsen (1967a)
Bobwhite quail (*Colinus virginianus*)	Maximum NaCl solution		605	Na: 284	NcNabb (1969b)
California quail (*Lophortyx californicus*)	Cloacal		669	Na: 365	NcNabb (1969b)
Gambel's quail	Samples = max. single values		669	Na: 470	NcNabb (1969b)
California quail	Max. NaCl solution		751	Na: 322	Carey and Morton (1971)
Gambel's quail	Cloacal samples		962	Na: 493	Carey and Morton (1971)
				Cl: 363	Smyth and Bartholomew (1966b)

Table 5.5 (continued)

Species	Notes	Osmotic urine-to-plasma ratio	Osmotic and ionic concentration		References
			mOs	mequiv/l	
Australian pipit (*Anthus novaeseelandiae*)	Cloacal urine	3 (for Na)	—	Na: 570	Rounsevell (1970)
Black-throated sparrow (*Amphispiza bilineata*)	Max. single cloacal value	—	—	Cl: 703	Smyth and Bartholomew (1966a)
Rock wren (*Salpinctes obsoletus*)	Max. single cloacal value			Cl: 403	Smyth and Bartholomew (1966a)
Sage sparrow (*Amphispiza belli*)	Cloacal samples	1.9	674	Cl: 345	Moldenhauer and Wiens (1970)
Brown-headed cowbird (*Molothrus ater*)	Cloacal samples	2.5 (for Cl)		Cl: 380	Lustick (1970)
Domestic turkey			553	Na: 227	Skadhauge and Schmidt-Nielsen (1967a)
Cormorant (*Phalacrocorax auritus auritus*)	Cloacal samples			Cl: 300	Schmidt-Nielsen et al. (1958)
Birds with salt glands:					
Herring gull (*Larus argentatus*)	Cloacal samples	—	463	Na: 105	Douglas (1970)
Black swan (*Cygnus atratus*)	Cloacal samples	—		Na: 266	Hughes (1976a)
Red-winged blackbird (*Agelaius phoeniceus*)	Cloacal samples	—		Cl: 289	Hesse and Lustick (1977)
Domestic duck	Seawater load	1.4	444	Cl: 124	Holmes et al. (1968)
Domestic duck				Cl: 162	Scothorne (1959)
			462	Na: 133	Skadhauge and Schmidt-Nielsen (1967a)
Herring gull (*Larus argentatus*)	Ureteral samples, salt load, single bird		525	Cl: 126	Holmes (1965)
			736	Na: 66	Douglas (1970)
Glaucous-winged gull (*Larus glaucescens*)	Cloacal samples	2.0	668	Cl: 58	Hughes (1977)
Jackass penguin (*Spheniscus demersus*)	Control	2.1	651	Cl: 33	Erasmus (1978a)
	Salt load	1.9	635	Cl: 114	Erasmus (1978a)
	Anaesthetised	2.2	651	Cl: 110	Oelofsen (1973)

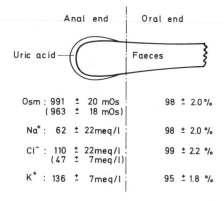

Fig. 5.4. Comparison of anal and oral ends of zebra finch droppings. Osmolality and electrolyte concentrations in oral end are presented as percentage of those in anal end. The birds received 0.6–0.8 M NaCl as drinking fluid. *Numbers in parantheses* indicate observations on dehydrated birds. The difference between oral and anal end, which is covered with precipitated uric acid and urates, appears to be small. Reproduced from Skadhauge and Bradshaw (1974)

	Anal end	Oral end
Osm :	991 ± 20 mOs	98 ± 2.0 %
	(963 ± 18 mOs)	
Na⁺ :	62 ± 22meq/l	98 ± 2.0 %
Cl⁻ :	110 ± 22meq/l	99 ± 2.2 %
	(47 ± 7meq/l)	
K⁺ :	136 ± 7meq/l	95 ± 1.8 %

The maximal osmotic urine-to-plasma ratio has been determined in only a few species by direct collection of ureteral urine. In most cases either the anal part of natural droppings (Fig. 5.4) after collection under oil, mostly in granivorous birds, or the liquid milky-white part of the droppings, mostly in carnivorous birds, has been centrifuged, and osmolality and solute concentrations measured on the supernatant. In the domestic fowl (Skadhauge 1968) and in the budgerigar (Krag and Skadhauge 1972) the anal ends of the droppings have almost the same osmotic concentration as ureteral urine, so the error induced by taking the composition of these parts of the natural droppings as equivalent to ureteral urine would pressumably be small. One exception to this generalisation is the possible release of trapped cations from uric acid precipitates in the cloaca (see p. 90). In the zebra finch a systematic comparison of the lower (anal) versus the upper (oral) end of the faeces "drumsticks" (see Fig. 5.4) showed the osmolality and Na, K, and Cl concentrations of the upper end to be within 95% of the concentrations of the lower end.

There is, however, another source of potential error. After large oral salt loads some fluid may have passed rapidly through the intestinal tract as observed in the domestic goose (Peaker and Linzell 1975; see p. 135) and in the herring gull (Douglas 1970). The cloacal fluid would in these cases not represent renal urine, but rather a mixture. Finally, some effect of ion absorption by the cloaca is apparent when the oral end of the droppings had 10%–30% lower concentrations of Na or Cl (Skadhauge 1974a) and natural voided samples had lower ion concentrations than samples of ureteral urine (Bindslev and Skadhauge 1971b; Hughes 1970a). Osmotic ratios of urine to plasma and the urine osmolalities have been summarised in Table 5.5, together with the concentrations of strong electrolytes for dehydrated, and salt-loaded birds without and with salt glands. Inspection of this table shows first the limited concentrating ability of birds. With the exception of two Californian salt-marsh sparrows, a higher concentrating ability than 1000 mOs corresponding to a urine to plasma osmotic ratio of 3, has not been observed. Second, in the dehydrated state only a small fraction of total osmotic space (of the supernatant) is occupied by the strong electrolytes (see also Table 5.6). Third, with the exception of the salt-marsh sparrows, there is not a strict correlation between concentrating ability and availability of water in the habitat, exposure to salt intake, etc. This is

Table 5.6. Concentration of several ions in supernatant of birds urine (unit: mM/l)

Species	Notes	Na	Cl	K	Mg	Ca	NH$_4$	PO$_4$	Osm	References
Domestic fowl	Wheat + barley	41	36	73	13	7	120	130	582	Skadhauge (1977)
	Commercial food	141	121	62	11	11	50	132	594	
	Control	33	56	116	—	—	90	107	—	Sykes (1971)
	Plant protein	31	41	109	12	3	53	48	—	Bokori et al. (1965)
	Animal protein	57	48	21	5	4	50	36	—	
Domestic turkey	Dehydrated	102	108	67	—	1	—	—	517	Vogel et al. (1965)
Domestic duck	Fresh water	9	12	37	—	1	233	39	241	Holmes et al. (1968)
	Sea-water	76	124	46	—	4	212	45	444	
Herring gull (*Larus argentatus*)	Salt loading[a] (early)	105	—	15	33	6	66	—	463	Douglas (1970)
	Salt loading[a] (late)	24	—	13	83	16	164	—	500	
Galah (*Cacatua roseicapillus*)	Dehydration	78	51	30	—	—	124	118	618	Skadhauge and Dawson (1980a)
Emu (*Dromaius novae-hollandiae*)	Disturbance diuresis	66	78	31	—	—	13	2	223	Skadhauge et al. (1980)
Ostrich (*Struthio camelus*)	Control	85	158	266	—	1	50[b]	3	—	Schütte (1973)

[a] Cloacal urine
[b] Urea + NH$_4$

70

obviously due to the fact that it is not renal concentrating ability alone that determines osmotic adaptation in birds, and that the cloaca and the salt gland must be taken into account (see Chap. 6, 7, and 8). If in particular Australian species investigated by Skadhauge (1974a) are considered, there clearly seems to be a rather narrow range for concentrating ability within each family, but the xerophilic birds differ considerably. Note the large difference between the emu and the ostrich. This may speak against a close systematic relationship between these two ratite species. Concerning adaptation to salt loading among birds without salt glands, only the Californian salt-marsh sparrows have the renal capability to cope with concentrated solutions. Species with salt glands do not concentrate their urine more effectively than others.

5.1.8 Renal Excretion of Chloride and Potassium Ions

The fraction of filtered amount of a substance that is excreted in ureteral urine is a precise measurement of the overall handling of that substance in the kidney. In only a few investigations on birds have the concentrations of both a filtration marker, and either Cl or K ions or both, been measured in urine and plasma. In several other cases, however, Cl and K have been analysed in urine together with Na. From the concentration ratios in urine and a fairly precise knowledge of fractional Na excretion (see Table 5.4) and ionic concentrations of plasma (see Table 1.4) the fractional excretion of Cl and K can be estimated. Inspection of Table 5.5 shows that the relative concentrations of Na, Cl, and K in ureteral and cloacal urine is close to 1:1:0.5–2. Only in certain cases of concentrated urines of dehydrated birds, particularly the Australian birds observed by Skadhauge (1974b), does the relative concentration of K exceed that of Na and Cl by a factor of as much as about 10. These observations indicate that the amount of Cl excreted is, as for Na, about 1%, whereas 15%–60% of filtered K is excreted in ureteral urine. In the concentrated urines the excretion of K will presumably exceed the filtered amount, indicating tubular secretion of this ion, whereas the Cl ion is largely resorbed, presumably together with the Na ion in identical parts of the nephron.

In a few investigations a detailed analysis of the renal handling of Cl and K have been carried out. The excretion of Cl is very close to that of Na in the domestic fowl, budgerigar and domestic duck (see Table 5.3). The excretion of filtered K varies from 13% to 60% in experiments in the domestic fowl, domestic duck, European starling and Gambel's quail (see Table 5.3). Tubular secretion of K was clearly demonstrated in an infusion study in the domestic fowl, in which Orloff and Davidson (1959) observed an increase in fractional excretion of filtered K from 18% in the control state to 312% following infusion of K_2SO_4 into the leg vein. This demonstrates directly tubular secretion of K.

5.1.9 Concentration of Several Ions in Ureteral Urine

Measurement of concentrations of several ions on the same samples of ureteral urine have been compiled in Table 5.6. Single determinations of ions measured less frequently, such as SO_4, have been included. Although urine osmolality was not

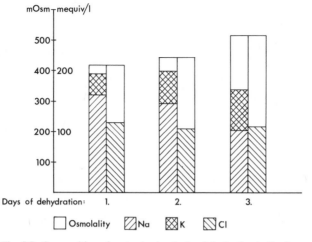

Fig. 5.5. Composition of ureteral urine during dehydration in the domestic turkey. During three days of dehydration urine osmolalities increase and the concentrations of Na and K are reduced and increased, respectively, but the total fraction of osmotic space occupied by "strong" electrolytes is reduced. Based on data from Vogel et al. (1965)

measured in all studies, inspection of Table 5.6 shows some relations very clearly: The majority of the "osmotic space" is—unless the NaCl intake is high—made up of potassium- ammonium-phosphate with each of these ions responsible for around one-quarter of total solute concentration. The NaCl concentration varies from 20% to 50% depending on the salt intake. Divalent ions, Mg, Ca, and SO_4, constitute around 5% each, urea and HCO_3 less. The high concentration of the ammonium ion, often reaching 200–300 mequiv/l, is remarkable and is related to the acid-base excretion in urine (Wolbach 1955; Skadhauge 1977).

The main conclusion from these observations is that the fluid that reaches the cloaca is very different from that encountered higher in the intestinal tract, and that it is of extensively variable composition. This has been taken into account in recent in-vivo perfusion and in-vitro studies of cloacal transport (see Shap. 6).

It is interesting to note the large fraction of total urine osmolality contributed by the measured ions and other solutes. A fraction of 74% was determined taking approximate osmotic coefficients into account for domestic fowl receiving a normal commercial diet (Skadhauge 1977). Amino acids attain a total concentration of only about 2 mM/l (Boorman 1971; Boorman and Falconer 1972).

Two osmoregulatory problems are of interest in this context. First, whether progressive dehydration results in any change in ionic composition beyond the mere subtraction of water, and second, whether under lack of salt other ions substitute for Na and Cl. In other vertebrates Na/NH_4 and Cl/HCO_3 exchange processes across limiting ephithelia have been observed (Schmidt-Nielsen and Skadhauge 1967). However, the low concentrations of HCO_3 in urine of the dehydrated domestic fowl on a low Na diet (Skadhauge 1977) speak against the presence of such mechanisms in birds. Progressive dehydration often resulted in declines of concentrations of Na and Cl (Fig. 5.5), as observed in the domestic turkey (Vogel et al. 1965) and in the budgerigar (Krag and Skadhauge 1972), but this may reflect decreasing food intake, possibly in association with mineralo-corticoid stimulation. The increasing fraction of the osmolality not accounted for by the strong electrolytes is due to ammonium phosphate.

5.1.10 The Basis of Hypertonicity in the Medullary Cones

Hypertonicity of the medullary cones has been observed in four species, domestic fowl, domestic turkey (Skadhauge and Schmidt-Nielsen 1967b), budgerigar, and savannah sparrow (*Passerculus sandwichensis rostratus*; Emery at al. 1972). Skadhauge and Schmidt-Nielsen (1967b) froze kidneys of dehydrated, salt-loaded, and hydrated roosters and turkeys. The medullary cones were dissected out at $-18°C$, and the osmolality and ionic and urea concentrations measured after boiling the tissue. Although an intramedullary gradient towards the tip of the papilla was observed in the turkey, the mean differences between cortex and medulla were small. In dehydrated and salt-loaded roosters the increase in NaCl concentration is only about 16 mM/l. In one dehydrated turkey it was 30 mequiv/l Na, and 41 mequiv/l Cl. K was not concentrated in the medullary cone, and urea constituted less than 0.5% of the osmolality. Hydration obliterated the gradient. Emery et al. (1972) observed by microcryoscopy higher osmolalities in the medullary cone. In dehydrated and vasopressin-injected budgerigars the osmolality in the collecting duct ranged from 1200 to 1400 mOs. In *P. s. rostratus* the maximal collecting duct values ranged from 1300 to 1500 mOs. The authors observed a "plateauing" of the osmolality in the middle of the length of the medullary cone, and—most clearly in the *P. S. rostratus*—even a decline towards the tip. The vasa recta were found slightly hypo-osmotic compared with the loops of Henle and the collecting ducts.

The conclusion drawn from these papers is that the avian kidney, like the mammalian, has a counter-current multiplier system which seems to function on NaCl transport out of the thick ascending limb of the loop of Henle. It functions without contribution of urea which is most important in mammals. This difference may explain the decline in osmolality towards the tip.

5.1.11 Correlation of Structure and Function

In the previous section the positive correlation between relative papillary length and maximal osmotic urine to plasma ratio as observed in mammals was cited (see p. 65). Furthermore, for closely related species the papillary width has been correlated with the capacity for solute excretion at maximal concentration (Schmidt-Nielsen 1964). Both correlations are intuitively reasonable. The longer the counter-current multiplier nephron system is, the higher the concentrations that can be achieved by the active NaCl transport out of the ascending limb of the loops of Henle. This is equivalent to an electrical circuit in which larger voltage is obtained with batteries in series. The larger transport capacity attained in a relatively wider renal papilla is analogous to the higher current, but unchanged voltage, obtained with batteries in parallel.

The measurement in recent years of a fairly high concentrating ability in several species of birds (Table 5.5) and a simultaneous increase in number of anatomical studies have made it possible to consider whether the correlations mentioned above exist in birds. The structure of the medullary cone of the avian kidney has been described on p. 56. In this section the quantitative relations will be discussed.

Measurements of total weight of the avian kidney have been reported by Johnson (1968) and Hughes (1970c). Johnson summarised renal weights of 181 species from 20 orders. The kidneys weighed, as a rule, close to 1% of the body weight with a range from 0.6% to 2.1%. Kidneys from relatively small species generally exceed 1%, the reverse being true for larger species. Johnson found no correlation with habitat or types of food ingested. Inspection of his table also fails to suggest any relation to renal concentrating ability. Hughes (1970c) examined relative kidney weight in 51 non-passerine species with salt glands and 52 species without, in an attempt to evaluate the effect of this accessory osmoregulatory organ on kidney size, and concluded that its presence does not exert a sparing effect on the amount of renal tissue. On the contrary, the presence of salt glands was associated with a slightly higher weight per kg body weight of the kidneys.

Johnson and Mugaas (1972a, b), in investigations of several quantitative relations among selected species by light microscopy and by injection and clearing studies, found that the diameters of cortical and medullary lobules, and loops of Henle were not related to renal concentrating ability, nor was the number of collecting ducts entering each lobule (Johnson and Mugaas 1972a). When, however, the fraction of renal tissue occupied by medullary cones was related to the concentrating ability expressed as maximal salinity that the birds could drink without weight loss, a proportionality was found. Only 7.2% medullary tissue was present in the song sparrow (*Melospiza melodia juddi*) as compared with 22.2% in *Passerculus sandwichensis beldingi* (Johnson and Mugaas 1972b). In the species in which osmolality has been measured, the pattern was the same. A similar relation was observed in the relative amount of medullary cone tissue, which, in the effective concentrators observed was twice to three times greater. One, less general observation, was similar to the pattern in mammals. In two good concentrators, the zebra finch and the black-throated sparrow (*Amphispiza bilineata*), the medullary cones are long in relation to the size of the kidney.

The importance of relative amount of medullary tissue was stressed by Johnson and Ohmart (1973a) who, in several finches and sparrows, found an inverse correlation between concentrating ability and the weight of cortex per medullary lobule. In a paper on the large-billed savannah sparrow, Johnson and Ohmart (1973b) interpret the fact that a higher concentrating ability is associated with a larger fraction of medullary tissue to indicate that the good concentrators have a higher fractional population of nephrons with loops of Henle. This highly logical interpretation explains the lack of correlation with papillary length in birds. It is in agreement with the occurrence of reptilian-type and mammalian-type nephrons (see p. 64). In later work Johnson (1974) extended the analogy with mammals and introduced the concept of relative length of medullary cone as mean length of medullary cone multiplied by ten and divided by cube root of the kidney volume.

This parameter, for terrestrial species, is fairly well correlated to concentrating ability, but at least one marine species with a salt gland, the doubled-crested cormorant which concentrates fairly well—according to the Cl concentration of cloacal fluid, 300 mequiv/l (Schmidt-Nielsen et al. 1958)—was completely off the scale. Thus other factors are involved in the determination of the concentrating ability. For the Australian species, for which concentrating ability (Skadhauge 1974a) and renal microstructure (Johnson and Skadhauge 1975) have been investigated, a linear correlation was found between relative medullary length and fractional amount of medullary tissue (Skadhauge 1976b).

A different quantification of relative abundance of nephrons with loops of Henle was used by Poulson (1965) and McNabb (1969b), who expressed this as the "mean number of medullary lobule units per kidney cross section" or "the average number of medullary lobules in each kidney lobe". They observed a linear relation with osmotic ratio of urine to plasma in three species of quail (McNabb 1969b), the house finch, and two races of savannah sparrow (Poulsen 1965). No correlation with average length of loops of Henle was apparent.

In summary, renal concentrating ability seems to be predominantly related to relative amount of medullary tissue. This has, although not proven, been taken to indicate in good concentrators the presence of larger fraction of nephrons with loops of Henle, i.e. mammalian-type nephrons, as compared with reptilian-type nephrons. This would lead to higher concentrating ability as the relative transport capacity of the ascending loops of Henle is augmented.

5.2 Action of Antidiuretic Hormones

The response of the avian kidney to water loading and to dehydration suggests mediation via a neurohypophyseal, antidiuretic hormone. This mechanism, which is closely analogous to the regulation of water balance in mammals, is now well established. An octapeptide, arginine vasotocin (AVT), has been isolated from the neurohypophysis of birds; it is depleted from the neurohypophysis when birds are dehydrated, and its main action on the kidney is to reduce the water output. Furthermore, the total quantitative and qualitative relations prove, beyond reasonable doubt, that AVT is a true hormone, i.e. participates during normal conditions in the regulation of water balance.

AVT is an octapeptide with a ring of five amino acids, linked through the disulphide bond of cystine, and a side chain of three amino acids. It occurs normally in all lower vertebrates (Sawyer 1967). The peptide has the ring structure of oxytocin and the side chain of arginine vasopressin (AVP), the neurohypophyseal hormone of most mammals (Table 5.7). It is interesting that this hormone was synthesised by Katsoyannis and du Vigneaud (1958) before it was actually known to occur in nature. Shortly thereafter, a polypeptide was isolated from the neurohypophysis of birds (Munsick et al. 1960; Munsick 1964) which was shown by 11 different bioassays to be identical to synthetic AVT. In agreement with its molecular composition, it causes some effects similar to those of oxytocin, such as

Table 5.7. The chemical structure of arginine vasotocin (AVT), arginine vasopressin (AVP), and oxytocin. These octapeptides have a ring of five amino acids and a side of three amino acids. AVT deviates only in one amino acid from the other hormones, from oxytocin in the side group, from AVP in the ring

---

Arginine vasotocin

Cys-Tyr-Ile-Gln-Asn-Cys-Pro-Arg-Gly

Arginine vasopressin

Cys-Tyr-Phe-Gln-Asn-Cys-Pro-Arg-Gly

Oxytocin

Cys-Tyr-Ile-Gln-Asn-Cys-Pro-Leu-Gly

---

contraction of the oviduct, and some similar to the action of AVP, such as antidiuresis in the domestic fowl. The identity of the antidiuretic principle of neurohypophyseal extracts of birds and synthetic AVT was also shown by chemical identification (Chauvet et al. 1960; Acher 1963). In the following section the identification of the hormone in the neurohypophysis of birds is discussed together with the symptoms induced by ablation of the neurohypophysis and the occurrence of spontaneous diabetes insipidus in birds. The main emphasis will be on the action of AVT on the kidney. Extrarenal activities of the hormone will only be mentioned briefly.

Not only AVT, but also mammalian hormones exert an antidiuretic action in birds; the difference is largely quantitative (Skadhauge 1968). Early observations on renal effects of mammalian neurohypophyseal hormones and extracts are summarised below.

5.2.1 Investigations of Mammalian Hormones

Burgess et al. (1933) injected Pitressin (Parke-Davis) intramuscularly into hydrated domestic fowl, following which urine volume decreased by 40%–90%. Korr (1939) injected Pitressin into one hen and observed an increase in urine osmolality from approximately 90 mOs to 490 mOs. Holmes and Adams (1963) gave Pitressin to domestic ducks. The urine volume was reduced by approximately 50%. Vogel et al. (1965) infused AVP into hydrated domestic turkeys weighing 3–5 kg. Infusion of 2–4 mU/kg (U = unit) did not result in any antidiuresis, a result to be expected since the mammalian hormones are quantitatively inferior to AVT. Skadhauge (1964) observed the mammalian hormone, the III international standard (ox vasopressin), to be less active than AVT by a factor of 10^4–10^5, weight for weight, in the hydrated domestic fowl as judged by the antidiuretic response. Lysine vasopressin, the antidiuretic hormone of pigs, was by a factor of at least 100 less active, as judged by antidiuretic activity, than AVT in the hydrated domestic fowl (Skadhauge 1968).

76

The activity range for the antidiuretic action of AVT in the hydrated domestic fowl was 5–50 ng/kg after intravenous injection. It can be calculated that 1.8 to 1.9 U of lysine vasopressin is equivalent to 20–40 ng AVT. Since lysine vasopressin contains 246 U/mg pure peptide, as judged by the rat antidiuresis assay (Berde and Boissonais 1968), the dose given was equivalent to 7000 ng lysine vasopressin. Since AVT has the same activity/mg, the 7000 ng lysine vasopressin can thus be considered equivalent to 24–40 ng AVT.

5.2.2 Amount of Arginine Vasotocin in the Neurohypophysis

Indirect estimates of hormone contents of hypothalamus and hypophysics have been made by histochemical methods. Farner and Oksche (1962) concluded that AVT is produced by hypothalamic neurosecretory cells, transported to the neuro-hypophysis with a stainable carrier and released under dehydration or osmotic stress. Oksche et al. (1959) observed that dehydration depleted the pars nervosa, but not the external zone eminentia of the white-crowned sparrow of stainable neurosecretory material. Similar findings were made in the zebra finch by Oksche et al. (1963) who observed that water restriction or salt loading made the neuro-secretory system more active with enlargement of the cells and occurence of neuro-secretory droplets. Again, however, the external zone of the eminentia medialis did not become depleted of neurosecretory material. Uemura (1964) observed that water deprivation for 7 days did not deplete the neurosecretory system of the bud-gerigar. Since these birds hardly lost weight, a longer period of water deprivation may be necessary in order to cause histochemical changes. In the domestic fowl Legait and Legait (1955), Graber and Nalbandov (1965), and Lawzewitsch and Sarrat (1970) found that the neurohypophysis was depleted of neurosecretory material after dehydration and salt loading. The response of the eminentia medialis was variable, presumably depending upon the length of the period of osmotic stress. Follett and Farner (1966) observed depletion of both neurosecretory material and neurohypophyseal hormones from eminentia medialis and pars nervosa of the Japanese quail after acute dehydration. About 80% reduction in AVT of pars nervosa was observed after 4 days of intake of hypertonic (0.2 M) saline solution. These authors found around 50 mU AVT/gland as determined by a specific pressor bioassay. Finally, Kripalani et al. (1967) studied the effects of 12 h of dehydration on the hypothalamo-neurohypophyseal system of three species of *Lonchura* of different climatic habitats. In these species dehydration produced the largest activation of neurosecretory nuclei and depletion of neurosecretory material from the neurosecretory tract and from the neurohypophysis in the species living in humid regions. The response declined gradually from the mesic to the xeric species.

The amount of AVT in the neurohypophysis of birds has been estimated in a number of species. The total content was estimated at only a few per cent of the amount that is necessary to induce conspicuous renal antidiuresis in hydrated birds. In the neurohypophysis of the domestic fowl estimates of vasoactive or antidiuretic material, most likely to be AVT, have been carried out by Heller and Pickering (1961) who dissected the "neuro-intermediate lobes" in domestic fowl and assayed by the rat blood-pressure method. A mean activity of 245 mU/gland was found. Since the weight of the birds was not stated, the content per kg can only be given approximately with 125–245 mU/kg. This can be converted to 500–1000 ng/kg since 243 mU of AVT corresponds to 1 μg of the pure AVT on the basis of the rat blood-pressure assay (Berde and Boissonais 1968; p. 815). The rat antidiuresis assay gave almost the same results. Ishii et al. (1962) measured the neurohypophyseal hormone activity of extracts of the eminentia medialis and pars nervosa of the pigeon in the normally hydrated state and after dehydration, by rat and frog-bladder bioassays. To control the specificity of the responses they used thioglycollate inactivation which breaks the disulphide bond in cystine in the ring. Dehydration reduced the content of neurohypophyseal hormone of pars nervosa to about one-eighth of the normal level, but no such change occurred in the eminentia medialis of the dehydrated pigeon. Hirano (1964) estimated the AVT content of the neural lobe of the hypophysis of the domestic fowl to be 112 mU/bird; the rat antidiuresis assay was used, controlled by inactivation with thioglycollate. In the domestic duck an average of 712 mU/bird was observed compared with 2.4 mU/bird in the budgerigar. If the ducks weighed 2–3 kg, the budgerigars 30–40 g, and the domestic fowl 1 kg, it would appear that no difference in AVT content was apparent among these species on a weight basis. In Munsick's (1964) experiments the neurohypophysis was dissected from roosters and assayed by rat blood-pressure method. This author found 1049 mg AVT/posterior lobe, approximately 1070 mU/kg equal to 4400 ng AVT/kg. Similar findings were made for the domestic turkey.

It appears that the neurohypophysis of the species investigated under normal hydration contains at least about 1000 ng AVT/kg body weight. As the physiological range of AVT acting on the kidney seems to be around 5–30 ng/kg body weight as single injection (see p. 80) the contents of antidiuretic material, undoubtedly AVT, of the posterior lobe, is 20- to 50-fold the renal dose. The conclusion is that the hypothalamo-neurohypophyseal system of birds produced enough AVT to regulate the water balance by excretion through the kidney. The observations of plasma disappearance of AVT in the domestic fowl (see p. 82) support this conclusion. It may safely be stated that AVT is a normal regulator of water balance in birds.

5.2.3 Diabetes Insipidus

The physiological role of neurohypophyseal hormones in avian antidiuresis is further strengthened by the occurrence of a diabetes insipidus-like condition after ablation of the neurohypophysis in the domestic fowl (Shirley and Nalbandov 1956) and the domestic duck (Bradley et al. 1971). Shirley and Nalbandov (1956) made selective neurohypophysectomy in immature hens with a similar number of birds left unoperated or exposed to a sham operation. Among these three groups no differences were observed after 4 weeks in body weight or weights of adrenal cortex, ovary, and oviduct. When the operation was repeated in adult egg-laying hens, the same egg-laying was observed in the three groups after a recovery period. The only dramatic difference was an increase in daily water consumption from

400–500 ml per bird in unoperated and sham-operated birds to 900–1000 ml per bird in neurohypophysectomised birds. In a few experiments the urine flow rate was measured; intramuscular injection of 0.5 U Pitressin reduced the urine flow rate by approximately 90%. Bradley et al. (1971) neurohypophysectomised young domestic ducks and compared effects with unoperated and sham-operated controls. The response 4 days after neurohypophysectomy was a 5- to 11-fold increase in drinking rate, which declined over 14 days to three times normal intake. The mean osmolality of urine was 151 mOs. The effect AVP and of ATV given in equal amounts (rat-pressor units) was measured. The urine flow rate during 3 h after intramuscular injections was returned to normal by 0.1 U AVP and by 0.01 U AVT.

Polydipsia has also been observed in the domestic fowl following lesions in the hypothalamus involving nucleus supraopticus. For example, Ralph (1960) observed a 2- to 3-fold increase of the drinking rate after electrolytic lesions in the supraoptic regions in the hypothalamus in hens. Koike and Lepkovsky (1967) showed that small lesions in this region caused a substantial decrease in antidiuretic activity of the posterior lobe of the neurohypophysis, but no difference in plasma osmolality between lesioned and unlesioned domestic fowl. The average urine flow rate in the operated birds was high; 17.4 ml/kg·h with a mean osmolality as low as 29 mOs. These birds thus became very polyuric. In a wild species, the white-throated sparrow (*Zonotrichia albicollis*) hypothalamic lesions resulted in a 274% increase in the drinking rate (Kuenzel and Helms 1970).

Spontaneous polydipsia was found in a strain of domestic fowl (Dunson and Buss 1968), in which the kidney responded to intravenous injection of AVT; 10 µg AVT more than doubled the urine osmolality, but did not produce an isosmotic ureteral urine. The sensitivity of the kidney cannot, on the basis of these experiments, be judged to be less than in normal, hydrated birds. Further studies on these birds seem to indicate that the polydipsia is partly primary, and partly secondary to a decreased sensitivity to AVT, thus resembling the hereditary human nephrogenic diabetes insipidus (Dunson et al. 1972). This renal abnormality must, however, be judged as mild since restricted water intake did not result in loss of body weight and even total dehydration resulted in a daily body weight loss of only 2.2%, equal to that observed in normal domestic fowl. The ability to drink saline solutions of 150–200 mM NaCl was reduced compared to normal birds, and the antidiuretic response to AVP injections was also reduced. The antidiuretic activity of the neurohypophysis was smaller, but that of the urine in specimens from this strain (Benoff and Buss 1976). Finally, indication of a primary polydipsia was found in two species of Galápagos mocking birds (*Nesomimus parvulus* and *macdonaldi*) (Dunson 1970).

A disturbance antidiuresis, as known in mammals, was inferred by Skadhauge (1968) who observed an antidiuresis induced by vein puncture equal to the injection of 20–30 ng of AVT.

5.2.4 Renal Effects of Arginine Vasotocin

In this section the effects of injections of AVT on renal functions in hydrated birds will be reported. The measured parameters include urine osmolality and flow rate,

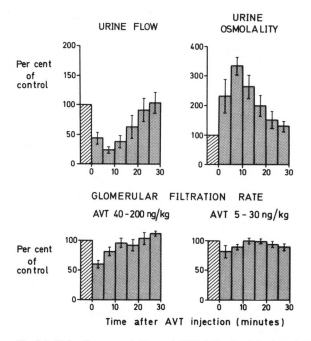

Fig. 5.6. Urine flow, osmolality, and GFR following injection of AVT. Values during antidiuresis are expressed as percentage of average value during control periods just prior to injection (*hatched area* 100%). Antidiuresis endures 30 min. The fractional change in urine flow and osmolality is nearly the same, resulting in almost unchanged rate of solute excretion. The GFR is reduced by higher doses of AVT. Reproduced from Ames et al. (1971)

GFR and filtration, tubular resorption, and excretion of Na, K, and Cl. Experiments have been performed on the domestic fowl, the domestic duck and the desert quail. The experiments had two main goals, demonstration of the extent to which AVT is able to reproduce the changes in renal functions caused by dehydration, and investigation of the mechanisms of action of AVT.

In three studies in hydrated domestic fowl doses of AVT given as single injections in the range of 5 to 30 ng/kg (Skadhauge 1964, 1968; Ames et al. 1971) resulted in antidiuresis for 20–30 min. The antidiuresis involved a pronounced reduction in urine flow rate with an almost proportional increase in urine osmolality. The net change in renal solute output was thus small. The results of 21 injections in eight birds (Ames et al. 1971) are summarised in Fig. 5.6. In these experiments the absolute values of urine flow rate decreased from 22.2 ml/kg·h to 5.3 ml/kg·h, i.e. to 23% of normal controls. The average osmolality increased from 102 mOs to 342 mOs and the osmotic ratio of urine to plasma increased from 0.33 to 1.11. This response is exactly identical to that observed in earlier experiments (Skadhauge 1964, 1968). It is also identical to the antidiuresis induced by AVP in mammals. In the hydrated domestic fowl a dose of about 30 ng AVT/kg thus produces the maximal urine osmolality after a single injection. In no case did single injections produce an antidiuresis as pronounced as in the dehydrated state. The dehydrated roosters of Skadhauge and Schmidt-Nielsen (1967a) had an average osmotic ratio of urine to plasma of 1.58, but the prolonged antidiuresis and very low flow rate observed after injection of several hundreds of ngs (Skadhauge 1968) makes it permissible to conclude that AVT is able to mimic the better part of a normal antidiuresis, probably under normal physiological conditions with a reduced extracellular fluid volume, the total antidiuresis.

The GFR was also measured as clearance of inulin or PEG in these experiments. Small doses of AVT, 5 to 30 ng/kg body weight, did not produce a significant decrease in GFR, whereas larger doses of 40 to 200 ng/kg body weight resulted in a reduction of 41%. After correction for errors in intrarenal and ureteral dead space the true decrease in GFR was calculated to be 20%. This appears close to the 23% decrease in GFR from hydration to dehydration in the domestic fowl (Skadhauge and Schmidt-Nielsen 1967a). The possible mechanisms in the decrease in GFR will be discussed below.

The total solute excretion calculated as the osmolar clearance was reduced by 30% during the hormone-induced antidiuresis (Skadhauge 1973; p. 31). The maximal reduction in Na and Cl output per unit GFR was also reduced by approximately 30% (see Fig. 5.6). The reduction was, however, of only 10 min duration. Rates of excretion of urea and K were not consistently changed.

It may be concluded that the main effect of AVT seems to be compatible with an increased osmotic permeability of the collecting tubules, whereas a large effect on tubular Na transport, as observed in lower vertebrates, is not apparent. The observations on the action of AVT in the domestic duck supports this conclusion. Bradley et al. (1971) observed a reduction of average rate of urine flow from 10.3 ml/kg·h to 3.6 ml/kg·h during the 3 h after injection of 41 ng AVT. The excretion of Na and Cl was, however, not reduced, whereas osmolar clearance and K excretion were both reduced by 44%. GFR was not measured.

The effects on renal functions and single-nephron GFR of injections of 10–200 ng AVT/kg have been investigated in Gambel's quail made diuretic with 2.5% of mannitol infused at a rate of 24 ml/kg·h (Braun and Dantzler 1974). A dose of 10 ng AVT/kg was clearly antidiuretic with a reduction of the urine flow rate to about 50%–60% of control rate, and an increase in urine osmolality from an osmotic ratio of urine-to-plasma of 0.8 to essentially isosmotic urine. The GFR was also decreased in the group receiving 200 ng AVT/kg to 60%–80% of the control level. The changes observed in the quail are thus close to those in the domestic fowl. The failure to produce a more hyperosmotic urine may be due to mannitol diuresis which is known in mammals to impair papillary hypertonicity.

An interesting aspect of the study of Braun and Dantzler was that of the effect of AVT on single-nephron GFR. As control values those observed in a previous study (Braun and Dantzler 1972; see p. 66) were used. At 10 ng AVT/kg both mammalian-type nephrons and reptilian-type nephrons had a GFR of approximately three-quarters of the control value, but only 52% of the reptilian-type nephrons were open as compared to 71% during control diuresis. This fraction was further reduced to 26% at 50 ng AVT/kg. The 200 ng AVT/kg dose resulted in two response types associated with a difference in the change in blood pressure. Reduction of blood pressure, or a biphasic response, was associated with reduced filtration by the mammalian-type nephrons and a reduced fraction of reptilian-type nephrons being open, whereas increased blood pressure was associated with total shutdown of reptilian-type nephrons, but a large increase of the filtration through mammalian-type nephrons. Although more work must be done to sort out the renal response when a change in blood pressure complicates the reaction, the main pattern: lability of the GFR, particularly of the reptilian-type nephrons as function of the AVT dose seems clear.

In conclusion, exogenous AVT appears to exert predominantly two effects on the avian kidney: (1) Constriction of the afferent arteriole to the reptilian-type nephrons (Braun 1976; see p. 65). This causes the overall decrease in GFR as observed during dehydration and salt loading. (2) Increase of the permeability of the collecting ducts of mammalian-type nephrons. This results in the characteristic antidiuresis with simultaneous increase in urine osmolality and decrease in flow rate. The combination of these effects augments renal concentrating ability (maximal osmotic ratio of urine to plasma) and reduces total solute and water output. The reduction of GFR aids in cloacal water absorption (see Chap. 8).

5.2.5 Metabolism of AVT, Plasma Concentrations, and Effects on Plasma Parameters

In this section estimates of plasma disappearance of AVT, measurements of plasma concentrations of AVT, and effects of injections of AVT on blood pressure (BP) and other parameters will be reviewed.

Hasan and Heller (1968) and Heller and Hasan (1968) have investigated the clearance of AVT from the blood of laying hens, non-laying hens, and cockerels. A large dose (5 U/kg) was injected intravenously and the blood concentration was measured at 10 min intervals up to 40 min after injection. The hormone concentration in plasma was estimated by the rat-pressor assay. This implies that the calculation of a unit is based on 255 U/mg pure peptide (Berde and Boissonais 1968). The dose used corresponds thus to approximately 20 μg/kg, which is 10^3 times larger than antidiuretic doses. The decay curve showed two exponential components in all three groups. The half-lives were 12–14 min in the first 20 min, 18–24 min in the following 20 min period. No clear difference was observed among the three groups. The authors interpret the steep initial decay curve to represent the "mixing" of the intravenously injected hormone. Since the free (non-protein bound) fraction of the hormone will be expected to leave the plasma and diffuse into the interstitial fluid at the start of the decay period and probably diffuse back to plasma when the AVT concentration is falling due to renal and hepatic clearance (Lauson 1967), a single exponential mean may perhaps best describe the "true" half-life. Thus a half-life of about 17 min can be calculated for the domestic fowl. This is a somewhat slower rate of inactivation than in mammals in which half-lives of 6–10 min have been found.

Measurements of plasma concentrations of AVT in the fowl have been effected by three groups (Sturkie and Lin 1966; Niezgoda and Rzasa 1971; Niezgoda 1975; Koike et al. 1979; Table 5.8). It appears from Table 5.8 that there is good agreement among the results obtained by frog-bladder bioassay (Sturkie and Lin 1966) and a radioimmunoassay (Koike et al. 1979). The latter authors state that storage of their

Table 5.8. Concentration of arginine vasotocin of the blood of four species

Species	Physiological state	AVT conc. ng/l ($\pm$SE)	References
Domestic fowl	Control, H_2O ad lib.	7$\pm$2	[b]
	Inj. of hyperosm. NaCl	23$\pm$12	
	During oviposition	934$\pm$150	
Domestic fowl	Control, H_2O ad lib.	661$\pm$92	[c]
Japanese quail		262$\pm$16	
Pigeon,		288$\pm$17	
Domestic fowl	Control, H_2O ad lib.	400$\pm$20	[d]
	Hyperosm. NaCl (420 mOs)[a]	1,100$\pm$500	
	Dehydration (418 mOs)[a]	3,300$\pm$1,000	
Domestic fowl	H_2O ad lib. (315 mOs)[a]	4$\pm$1	[e]
	Dehydration (356 mOs)[a]	24$\pm$3	

[a] Plasma osmolality
[b] Sturkie and Lin 1966 (Bioassay: Frog bladder permeability. Oxytocin standard. Conversion based on 360 U/mg)
[c] Niezgoda and Rzasa 1971 (Bioassay: Frog bladder permeability)
[d] Niezgoda 1975 (Bioassay: Frog bladder permeability)
[e] Koike et al. 1979 (Radioimmunoassay. Conversion based on 160 U/mg, rat-pressor assay)

samples may have reduced the hormone concentration. A resting level of approximately 10 ng/l plasma of AVT in non-laying domestic fowl with free access to water has thus been found. Egg-laying is accompanied by a manifold increase in concentration of AVT (see Table 5.8). Dehydration augmented the AVT concentration approximately sixfold (Koike et al. 1979). The assay of Niezgoda and Rzasa (1971) resulted in higher concentrations, but also observed was a three- to sixfold increase in plasma concentration of AVT during dehydration and after hyperosmotic NaCl loading (Niezgoda 1975).

In the domestic fowl, larger doses of AVT exert an effect on BP and other parameters, such as levels of blood sugar and plasma electrolytes. The latter effects can be considered to lie outside the field of osmoregulation. Among older observations Woolley (1959) observed a biphasic, but largely depressive effect on the emu, a cormorant (*Phalacrocorox varius*) and a penguin (*Eudyptula minor*) after injections of about 0.5 U/kg of Pitressin to birds anaesthetised by Na-phenobarbitone. The reduction of BP in the domestic fowl has even been proposed as a basis for a bioassay for posterior pituitary powder (Coon 1939). In recent experiments with use of AVT (Ames et el. 1971) a decrease in BP with doses less than 50 ng AVT/kg could not be detected. It may therefore be concluded that for the small doses of AVT, e.g. 30 ng/kg, that are sufficient to induce a maximal antidiuresis, a decline in BP does not form part of the physiological effects of the hormones observed in anaesthetised domestic fowl. Doses of 150–500 ng AVT/kg induced fall in BP for less than 5 min.

In Gambel's quail Braun and Dantzler (1974) observed three types of response in BP to injections of 200 ng AVT/kg, an increase in BP, a decrease, or a biphasic response. The deviation was within a range of $+20$ and -30 Torr. The average blood pressure was 112 Torr. It is concluded that effects on BP after injections of AVT seem to be largely pharmacological.

Finally, it should be noted that Rzasa and Niezgoda (1969) and Rzasa et al. (1974) observed an increase in concentration of Na in whole blood and plasma after 100–400 ng AVT/kg in the domestic fowl. The effects were, however, not proportional to the dose of AVT. A slight increase in blood volume and decrease in plasma protein concentration were observed after 100 ng AVT/kg.

5.3 Renal Nitrogen Excretion

Because of the low solubility of uric acid, uricotelism permits excretion or storage of nitrogen waste in a small volume of water. This is essential for development of the embryo in the egg of reptiles and birds. Uricotelism may also be viewed as pre-adaptation for water conservation. Homer Smith (1953; p. 123) calculated that the nitrogen formed from 1 g of protein would require 20 ml of water for excretion as urea in blood isosmotic solution. Birds and reptiles excrete this amount of nitrogen as uric acid in less than 1 ml of urine without an efficiently concentrating kidney.

In this section the renal excretion of nitrogenous constituents will be considered. First, the state of uric acid and urates in ureteral urines. Second, the renal clearance of uric acid and urates, and their total concentration in the ureteral urine as compared with the amount in solution. Third, the excretion of the various nitrogen compounds as fractions of total nitrogen excretion. Fourth, the possible change in the ratio of the various nitrogen constituents as function of the protein intake and the osmotic state of the bird. Fifth, the possible trapping of ions with the precipitates of uric acid and urates.

The aim is to quantify the role in osmoregulation of the excretion pattern, and the possible adaptations. It is relevant to point out both for birds and reptiles that the excretion of a majority of nitrogen waste as insoluble substances, either precipitated or in supersaturated colloid suspension, is a pre-requisite for storage of ureteral urine in a resorptive segment of the gut without absorption of the nitrogen. A short comparative survey of this question is given in the introduction to Chap. 6. Early investigations determined uric acid by direct colorimetry, whereas in later work specific enzymatic methods (by uricase) were used. The general picture has, however, not changed as a result of more precise analytical methods. Still standing is the conclusion by Young and Dreyer (1933): "The urine of the fowl is basically a supersaturated solution of uric acid or urate, part of which is in colloidal form." Several workers have investigated the conditions under which urates and uric acid form colloid suspensions. Young and Musgrave (1932) observed that optimal conditions exist for gel formation, i.e. for supersaturated solutions of uric acid and urates to remain in colloid suspension at the temperature and pH of ureteral urine (5–8.5 for Na-urate). Schade and Boden (1913) showed that formation of gels by supersaturated solutions of Na-urate is facilitated by higher concentrations of the solutes, particularly Na and K ions occurring in concentrated avian urine. This was confirmed by Young et al. (1933). Porter (1963a) concluded that supersaturated solutions of urates in phosphate buffers are colloidal, and Porter (1963b) confirmed that the cations of ureteral urine are important for the formation of flocculates of uric acid and urates.

5.3.1 Excretion of Uric Acid and Urates

The first study to report precise measurements of renal clearance of uric acid in a bird seems to be that of Shannon (1938b). The earlier literature is extensively covered in his paper and will, with few exceptions, only be quoted here in connection with relative excretion rates of nitrogenous constituents (Table 5.9). Shannon (1938b) measured renal clearance of uric acid and inulin simultaneously on conscious male domestic fowl. Tubular secretion of uric acid was proven by the observation of a uric acid/inulin clearance ratio ranging from 7.5 to 15.8 at a normal plasma uric acid level. The amount secreted, as compared to filtered, is 87%–93% of the total excretion of uric acid. Infusions of uric acid allowed the uric acid/inulin clearance ratio to be determined in the range of 0.3–6 mM/l uric acid in plasma. At the high plasma concentrations typical self-depression of clearance was observed. At 6 mM uric acid in plasma the clearance ratio was only 1.8–3.2 because the filtered rate is high compared with the tubular resorptive capacity. This allows determination of T_{max} for uric acid. Stewart et al. (1969) observed a uric acid/inulin clearance ratio of 6.6 in both fresh water and hypertonic saline maintained domestic ducks. The uric acid clearance was around 1 ml/kg min, equal to 90%–93% of the PAH clearance.

Recent information on tubular secretion of uric acid has been excellently summarised by Dantzler (1978). As the excretion of uric/acid depends on tubular secretion, it is largely independent of tubular water resorption (urine flow rate). The ureteral concentration of uric acid is consequently high on a low rate of urine flow (dehydration). Typical concentration values of 100–150 mM/l have been observed in the domestic fowl with a maximal number of 1.3 M/l observed by Gibbs (1929) as the highest. The values far surpass the solubility which is 0.38 mM/l for uric acid, 6.8 mM/l for Na-urate, and 12.1 mM/l for K-urate (Dantzler 1978). The amount of uric acid/urates in solution thus contributes only a few mM to the urine osmolality. The remainder present in supersaturated colloid suspension does not contribute measurably to osmotic pressure (Skadhauge 1973; p. 34) as it totals only around 10^{10} particles/l. This is the basis of water saving by uricotelism. It is also quite clear that the factors that keep uric acid in colloid suspension are of importance for the flow of urine through collecting ducts and ureters.

The extremely high concentrations of uric acid in ureteral urine of domestic fowl on a low rate of urine flow were confirmed in the domestic pigeon by McNabb and Poulson (1970) who measured a concentration of 1.1 M/l. They also measured the fraction of total uric acid precipitated by collection of ureteral urine at 42°C and centrifugation at that temperature at 500 g. This force is too low to precipitate colloids. They observed a near linear relation from 0 to 110 mM/l uric acid at which concentration around 90% of total uric acid was precipitated (Fig. 5.7). At that concentration at which a urine osmolality of 650 mOs was observed, nearly 40 mM/l of uric acid was not precipitated. As only 9 mM/l of the uric acid and

Table 5.9. Nitrogen excretion in avian urine

Species	Notes	Total N g/l	Uric acid urates	NH$_4$	Urea	Creatine-creatinine	Amino acids	Purines	References
Domestic fowl		1.3	85.9	1.5	1.0	—	2.5	1.7	Szalágyi and Kriwuscha (1914)
Domestic duck		2.2	77.9	3.2	4.2	—	2.7	0.5	
Domestic fowl		—	81.9	5.6	—	—	—	—	Katayama (1924)
Domestic fowl		10	62.9	17.3	10.4	8.0	—	—	Davis (1927)
Domestic fowl		—	65.8	7.6	6.5	4.6	—	9.6	Coulson and Hughes (1930)
Domestic fowl	Fed	—	81.8	10.9	2.4	2.4	—	—	Sell and Balloun (1961)
	Fasted	—	74.6	18.6	2.9	4.0	—	—	
Domestic fowl	Fed	4.4	84.1	6.8	5.2	0.5	1.7	—	Sykes (1960)
	Fasted	2.4	57.8	23.0	2.9	4.3	2.8	—	
Domestic fowl		—	80.7	10.5	4.5	0.9	2.2	—	O'Dell et al. (1960)
Domestic fowl	Low protein	—	58.7	29.4	3.4	6.8	—	—	Tasaki and Okumura (1964)
	High protein	—	79.2	14.9	1.2	0.9	—	—	
Domestic duck	Fresh water	6.4	51.9	30.3	—	1.3	—	—	Stewart et al. (1969)
	Hypertonic saline	11	56.8	27.5	—	2.5	—	—	
Domestic fowl	Low protein	11	54.7	17.3	7.7	—	—	—	McNabb and McNabb (1975)
	High protein	13	72.1	10.8	9.7	—	—	—	
Turkey vulture	Fed	61	87	9	4	—	—	—	McNabb et al. (1980)
(*Cathartes aura*)	Fasted	13	76	17	7	—	—	—	

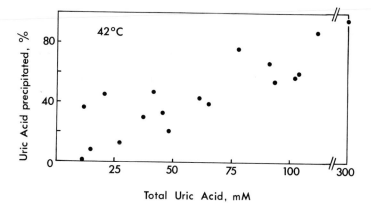

Fig. 5.7. Precipitation of uric acid a as a function of total concentration of uric acid in the pigeon. The proportion of precipitated uric acid in ureteral urine collected at body temperature increases approxymately linearly up to 90%, precipitated at a total uric acid concentration of 110 mM. Reproduced with permission from McNabb and Poulson (1970)

urates were calculated to be in solution at the temperature and pH of the solution, three-fourths of the amount must be in a colloidal state. The ratio between the amounts in solution, in colloidal state, and in precipitation was thus 1:2:12. It is interesting to note that if the maximal renal concentration of uric acid, which was ten times higher than in this experiment, was reached a urine osmolality of around 900 mOs would be required to excrete the nitrogen in the form of urea. In addition, the other solutes would need "osmotic space" for renal excretion.

The fractions of total urinary nitrogen excreted as uric acid urates, ammonium ion, urea, and other nitrogenous substances are recorded in Table 5.9. As realised already at the beginning of the nineteenth century (Brown 1970), uric acid and urates normally constitute 70%–80%, and ammonium ion 10%–20%, with smaller fractions excreted as urea, creatine-creatinine, amino acids, and purines. Most observations have been made on the domestic fowl, but measurements on the domestic duck show a similar pattern. Findings on the carnivorous turkey vulture (*Cathartes aura*) will be discussed on p. 90.

The fractional excretion of uric acid-urates is, as to be expected, augmented by feeding as compared to fasting, and by a high protein content of the diet. The accompanying decline in fractional ammonium excretion is only relative. The rate of NH_4 excretion is actually increased in fed and protein-loaded birds as both urine flow rates and NH_4 concentrations go up. No change in fractional excretion between insoluble and soluble nitrogen waste has been observed as a function of water availability (see p. 88). Furthermore, the concentration of NH_4 in the urine of birds on a low NaCl diet is high (see Table 5.5) presumably just because the NaCl ions are absent. The excretion of ammonium ion is only regulated to serve the acid-base balance. A theoretically possible shift of end point of nitrogen metabolism does not seem to play a role in conservation of water or NaCl. The fact that a higher nitrogen load leads to higher fractional excretion of the insoluble uric acid/urates indicates, however, that water is saved. This effect has been quantified (see p. 89).

In 1969 Folk noted that freshly formed bird droppings were not formed of crystalline uric acid, but amorphous concrements (spheres) of 2–8 μ in diameter. Based on X-ray analysis, he claimed that this material did not contain uric acid although he observed a peak at 0.322 nm typical for, but lacking other characteristic lines of uric acid. His arguments were challenged by Willoughby (1970) and by Poulson and McNabb (1970). The matter was settled by Lonsdale and Sutor (1971) who observed that the white droppings from budgerigar consisted largely of uric acid dihydrate. They further demonstrated the presence of cations bound to the spheres of uric acid dihydrate which were soluble in water and dilute acids. This question of trapping of ions within the spheres of uric acid dihydrate has been explored quantitatively by McNabb et al. (1973) in roosters fed a high or low protein diet while receiving either tap water or 1% NaCl solution. Ureteral urine was collected in fasted birds over 2 h. A maximal average Na concentration of 451 mequiv/l in the supernatant was observed in the low protein/salt-water group, and a maximal K-concentration of 110 mequiv/l in the high protein/freshwater group. Much higher concentrations of ions were observed in the precipitates; in the former group an average Na concentration of 1256 mequiv/l, in the latter a K concentration of 544 mequiv/l. The maximal average ion concentrations of the precipitate were observed in the low protein/tap-water group: Na: 1954 mequiv/l, K: 577 mequiv/l. The precipitate (after centrifugation) may thus contain high concentrations of cations. The maximal fraction of cations excreted in the precipitate was 34% of K in the high protein/salt-water group, 75% of Na in the high protein/tap-water group. As all these collections are of short duration in the postabsorptive period, the differences should be viewed with some caution. The general pattern is, however, highly important. It suggests that uricotelism allows not only non-osmotic excretion of nitrogen waste, but also permits excretion of a significant amount of Na and K with the urates without demand for "osmotic space". McNabb (1974), in extension of these experiments to include measurements of the concentration of NH_4 in the supernatant and precipitate, observed an interesting difference between the NH_4 ion on one hand and Na and K ions on the other. When a more concentrated urine was formed and more than 80% of the uric acid was precipitated (at room temperature), the precipitation of both Na and K increased to about 70%–80% (Fig. 5,8), whereas only 15%–20% of the excreted NH_4 was precipitated. McNabb suggests that this important trapping of Na and K proceeds by adsorption or electrostatic binding to the uric acid dihydrates rather than by formation of urates. His investigation confirms the importance of uric acid precipitation for excretion of alkali metal ions. The fate of these bound ions in the cloaca would thus seem to be a major problem in avian osmoregulation. NH_4 remains largely in solution. In the study of McNabb the maximal supernatant NH_4 concentration was about 150 mequiv/l.

The precipitation of divalent cations and Cl ions has also been examined by McNabb and McNabb (1977) who found in roosters receiving tap water and a low-protein diet that of the total excretion of Ca and Mg 32% and 24%, respectively, were in the precipitate, which, however, contained virtually no Cl (0.3%). Drinking of salt water reduced these fractions. In their paper physical techniques (differential thermal, and thermogravimetric analyses) were reported to demonstrate different "lines" produced by monobasic urate salts and the precipitates, suggesting a

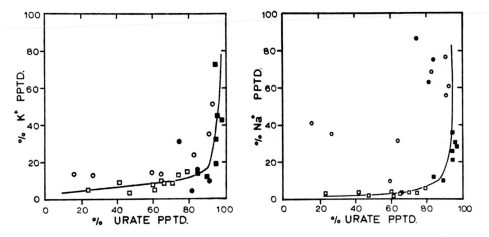

Fig. 5.8. Fraction of total urinary sodium and potassium in the precipitate as functions of urate precipitation in the rooster. The percentage of both Na and K in the precipitate remains less than 10% unless 80% of the total amount of uric acid is precipitated. In that case the fraction may rise to 70% of totally excreted cations in the precipitate. This only happens a high concentrations of arround 120 m M uric acid. Reproduced with permission from McNabb (1974)

different chemical composition. An interesting, but not yet quantifiable observation was a difference between the "lines" of ureteral urine and of naturally voided samples, which suggests a change in chemical composition after contact with the cloaca.

The material presented in the foregoing section confirms the water-conserving role of uric acid excretion as precipitate or in suspension, and demonstrates the role of "cation trapping" by uric acid spherules in the excretion of these ions. The next question is whether different rates of intake of protein and NaCl affect nitrogen excretion. The interesting problem is whether there is any adaptation that results in excretion of nitrogenous waste in a less toxic or less water-demanding form if the bird is stressed by a high intake of nitrogen or by a low water consumption. McNabb et al. (1972) studied total ureteral excretion of NH_4 in pigeons given high (44%) and low (11%) protein diets, and receiving either water ad libitum or the minimum water intake necessary to maintain stable body weight. The high protein diet resulted in a doubling of the water intake, and a four- to fivefold increase in NH_4 output. Water restriction led in both high and low protein diet groups to a doubling of the absolute rate of NH_4 excretion, and to fairly high average concentrations of NH_4 of ureteral urine, specifically 130 mequiv/l in the high protein group, 133 mequiv/l in the low protein group. It seems permissible to conclude that a greater fraction of total nitrogen was excreted as NH_4 during water restriction in the pigeon. Protein loading, however, did not change the rate of NH_4 excretion. Although the former may appear to be detrimental to water conservation, the change may serve other functions such as regulation of pH or conservation of Na. Ward et al. (1975a) made a similar study on male domestic fowl including measurements of excretion of uric acid and NH_4. In this species, also, water restriction resulted in a higher fraction of total nitrogen excreted as NH_4. The

89

mean urinary concentration of NH_4 was about 100 mequiv/l on restricted water intake. There was, however, no consistent change between the proportions of these nitrogenous compounds as functions of either dietary protein, nor changes in water availability (McNabb and McNabb 1975).

In a recent study McNabb et al. (1979) have examined the relationship of the availability of water and form of nitrogen excretion in a carnivorous species, the turkey vulture, either fasted or fed a commercial dog-food diet (12% crude protein). Water was available in both regimes, and identical urine osmolality was observed after natural voiding of liquid urine from the cloaca, i.e. 397 ± 15 mOs during fasting, $417 \pm$ mOs during feeding. The proportion of the various nitrogenous compounds was similar to the values observed on herbivorous birds during fasting (see Table 5.9), but feeding increased the fraction of uric acid/urate from 71% to 87%. The authors calculate a loss of 80 ml water/g N during fasting and only 16 ml water/g N after feeding. As the osmolality did not change, the authors suppose more cations are trapped in the urates or excreted through the salt gland during feeding. This study thus seems to demonstrate a water-conserving regulation of nitrogen excretion. Studies on other carnivorous birds are needed.

5.3.2 Quantitative Role of Cation Trapping in Precipitates of Uric Acid

The studies quoted in the previous section showed that the fraction of cations trapped in the uric acid precipitates varies according to the protein and salt content of the diet and, most significantly, according to the level of hydration. The range of fractional trapping was 3%–75% of Na, 8%–84% of K (see Fig 5.8) (and 17%–32% of Ca and Mg. In other studies higher as well as lower fractions have been found. Braun (1978) measured excretion rates of Na and K in the European starling during mannitol diuresis. The cation concentrations were measured in quantitatively collected fractions of urine before and after treatment with uricase (for analysis of uric acid). The concentration of Na and K increased from 14 and 23 mequiv/l to 966 and 144 mequiv/l on the average. The fractions of total amount of ions excreted, which was in association with the uric acid precipitates, was 98.5% of Na and 84% of K. This excretion was clearly not in the form of monobasic salts as the molar ratio of Na and K to uric acid were 16.5 and 2.1. Salt loading did not change these ratios. This study is in contrast to observations on dehydrated male domestic fowl (Skadhauge and Schmidt-Nielsen 1967a), in which urine from one ureter was washed out with a high rate of water flow for determination of total cation excretion, and urine from the other ureter was collected directly and the cation excretion rate determined as the concentration in the supernatant times the flow rate. The mean excretions of both Na and K determined on the supernatant without dilution were two-thirds of the total excretion. In other investigations (Krag and Skadhauge 1972) analysis of natural droppings after cloacal storage in two domestic fowl did not reveal significant trapping in the precipitate as compared to the supernatant. Long and Skadhauge (1980) also failed to detect cations trapped in the precipitates of ureteral urine of white Plymouth Rock hens. To the factors determining fractional ion trapping as outlined above must thus be added species difference and the possible influence of storage in the cloaca. Clearly, this field is most important for the comprehension of avian osmoregulation and must be investigated further.

5.3.3 Excretion of Urea

A priori urea would be expected to play a minor role in avian osmoregulation since its concentration in plasma is about 0.3 mM/l, and less than 10% of urinary, nitrogen is excreted in the form of urea (see p. 86). It is, however, interesting to note the extent to which filtered urea is resorbed in the bird kidney, and the possibility that urea significantly influences the formation of hypertonicity in the medullary cones. Concerning the first problem Pitts and Korr (1938) compared urea and inulin clearance in domestic fowl under infusion of external urea. The birds were in a state of hydration as the urinary flow rate was approximately 40 ml/kg·h. An average urea clearance of 1.50 ml/kg·min and an inulin clearance of 2.04 ml/kg·min were observed. The clearance ratio was thus 0.74, equivalent to a resorption of one-quarter of the filtered amount of urea. This finding was confirmed by Owen and Robinson (1964), who found an average urea to inulin clearance ratio of 0.7. After infusion of urea into a leg vein, excretion thereof increased in parallel from the ipsilateral and the contralateral kidneys, and a unilateral effect was not observed. A similar result was obtained by Sykes (1962), who also found equal concentrations of urea in urine from the two kidneys after unilateral infusion. The conclusion at this stage seems to be that the kidney of the domestic fowl handles urea just as the mammalian kidney. In spite of the renal portal circulation and a differently developed counter-current system, filtered urea is concentrated in the tubules and a back diffusion of around 30% occurs. Skadhauge and Schmidt-Nielsen (1967b) showed however, by measuring urea – inulin clearance ratio in highly hydrated and in dehydrated domestic fowl, that the ratio had a wide range in this species. In the hydrated state 100% of filtered urea was excreted, in the dehydrated state (inulin U/P ratio = 100) only around 1% of urea was excreted. A fraction of 99% was thus resorbed in the tubules. This resorption may occur because a much smaller accumulation of urea occurs in the medullary cones of birds than in the mammalian papilla. Stewart et al. (1969) observed in fresh water and hypertonic saline-maintained domestic ducks urea/inulin clearance ratios of unity and 0.3, respectively. This difference most likely is due to a higher flow rate of urine in freshwater-maintained birds. Skadhauge and Schmidt-Nielsen (1967b) observed in dehydrated domestic turkeys that concentration of the medullary cone urea increased by maximally 60% compared to the cortical concentration. In mammals a 30-fold increase is common. Urea does therefore not contribute significantly to medullary hypertonicity in domestic fowl or turkeys. In agreement with these findings Dicker et al. (1966) found little effect of urea infusion on either urine flow rate or solute excretion in the domestic fowl.

Chapter 6

Function of the Cloaca

6.1 General Aspects of Cloacal Function

Post-renal modifications of the urine play very different roles among the major vertebrate groups. Generally, more important changes in the urine occur in lower tetrapods: amphibians and reptiles, but also in some fish urine is modified in the bladder. In mammals there are only small changes. Birds appear to occupy an intermediate position (Skadhauge 1977).

Mammals regulate the body content of salt and water almost exclusively through excretion by the kidney. In the dehydrated state more than a slight permeability to urea or water of a post-renal resorptive epithelium would jeopardise nitrogen excretion and conservation of water by the kidney since urine, for example in man, may reach an osmolality 5 times that of plasma, and a urea concentration 100 times that of plasma. Fellows and Turnbull (1971), who have summarised the investigations of the permeability of the epithelium of the mammalian urinary bladder estimate that a dehydrated man may resorb less than 0.5% of renally excreted urea and Na in the bladder, but a 10% increase of the water content of the bladder may occur.

Birds are in a somewhat different position, although they also are able to make urine hyperosmotic to plasma, because the chief end-product of nitrogen metabolism is uric acid and urates. These, forming a supersaturated colloid suspension, are excreted without contributing significantly to osmotic pressure of the urine (see p. 85). Secondly, uric acid is not readily broken down to ammonia, and may therefore be stored in the intestine.

The mechanisms of modification of urine during storage in the cloaca are the subject of this chapter. The cloaca cannot concentrate the urine above the osmolality attained by the kidney, or even maintain this osmolality if this is higher than that of plasma, but some salt and water are resorbed. The cloaca may thus save a fraction of the ureterally voided salt and water in the dehydrated state. It may counteract the water and salt excretion in water-loaded and the salt-loaded states, but only slightly. It is relevant to point out that the result of the difference from the mammal is such that the best adapted species in the two vertebrate groups achieve the same independence of water: desert rats (*Dipodomys merriami*) (Schmidt-Nielsen et al. 1948), for example, can live without water on a diet of dry seeds alone, if the ambient temperature and humidity are not too unfavorable. The same ability has been observed in several xerophilic seed-eaters (see Table 2.2), among them the budgerigar and the zebra finch (see Chap. 10). The renal concentrating ability is, however, very different: An osmotic urine/plasma ratio of 14 was observed in the desert rat as compared to 3 in the budgerigar and the zebra finch (see Table 5.5).

6.1.1 Anatomy of the Cloaca

Urine moves in birds backwards from the urodeum into the coprodeum, large intestine, and even into the caeca (see Chap. 3). Therefore it is relevant to inquire into the type of epithelium in the cloaca and the parts of the gut with which the urine comes into contact. Obviously, any considerable exchange of ions or water between plasma and lumen will only occur across a resorptive epithelium. A macroscopic description of the intestinal tract of 17 orders of birds was given by Gadow (1879a), who measured length of caeca, and distance from anus along the cloaca and large intestine (Enddarm) to the openings of the caeca. Concerning the relative size of coprodeum versus large intestine, he noted that the coprodeum was particularly large in birds of prey and in ratites. In the latter group two unusual observations were made. A persisting dorsal bladder, 8 cm in length, 4.5 cm in width, was observed in the American rhea Nandu, and unusually large caeca and a very long rectum (large intestine or colon) were observed in an adult ostrich: The caecum was 70 cm in length, the large intestine 7–8 m, and the cloaca 20 cm. He concluded that the large intestine of the ostrich functioned similarly to the mammalian colon. Gadow had not access to an emu, but the 15–17 kg chicks investigated by Skadhauge et al. 1980 (see p. 95) had in comparison relativelly smaller coprodeum and large intestinal segments, although large absorptive areas. Gadow (1879b) tried to relate intestinal development to type of food: concerning the caeca he found as did other authors (see p. 32), no consistent picture, but he arrived at the same general conclusion: "Die Ausbildung der Blinddärme steht in direktem Verhältnis zur Menge der vegetabilischen (Leguminesen) Nahrung".

A histological account of the epithelium of the lower gut of birds was given by Greschick (1912), who noted in the large intestine, as well as in the coprodeum, a folded epithelium with irregular villi. The epithelium had columnar cells, some goblet cells, and crypts of Lieberkühn. The thorough description by Clara (1926) of the histology of the cloaca in a number of species noted that the epithelium changed from a single-layer columnar-cell epithelium in the urodeum to a multilayer squamous-cell epithelium in the proctodeum.

There is thus anatomical basis for resorption oral to the urodeum, i.e. in the coprodeum and large intestine (rectum or colon). The epithelial surface of the coprodeum and large intestine is enlarged by the presence of large irregular villi (Müller 1922). These are quite differently developed among different species but, as for example in the capercailzie (Clara 1926), the villi waith narrow intervillous spaces constitute a 1 mm thick layer.

A note on the site of storage of urine and faeces in the intestinal tract is appropriate in this section. In all the species examined by the author the proctodeum and urodeum were found empty of urine and faeces, and closed off from the coprodeum by a sphincter. Another sphincter closes the large intestine off at the junction of the caeca with the large intestine. There is no functional separation between coprodeum and large intestine so that the coprodeum and large intestine form one cylindroid compartment containing urine and faeces, this being only more distendable in the lower (coprodeum) than in the upper part (large intestine). Clara (1926) observed that the coprodeum was particularly wide and barrel-shaped in species that void very liquid excreta, such as birds of prey and

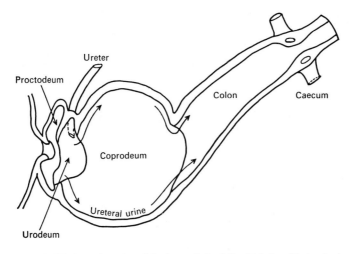

Fig. 6.1. The lower intestine of the domestic fowl. Sagittal view. Ureteral urine moves retrogradely from urodeum into coprodeum, colon (rectum, large intestine) and caecum. The uric acid is largely deposited in the coprodeum around a central cone in faeces. Reproduced from Choshniak et al. (1977)

herons. In the X-ray studies of the cloaca of the domestic fowl the coprodeum appeared almost spherical (Akester et al. 1967). A sagittal view of the lower gut in the domestic fowl is presented in Fig. 6.1. The retrograde movement of ureteral urine is treated in detail in the next section (see p. 97). The histology of coprodeum and colon of a number of Australian birds has been investigated by Johnson and Skadhauge (1975). They confirm the presence of simple columnar epithelial cells (with numerous goblet cells) lining colon and coprodeum. Crypts of Lieberkühn were present in both. They were deepest in colon and became progressively more shallow in the anal direction in the coprodeum. Four mucosal patterns were found: (1) zebra finches have well developed, regularly spaced, tall villi in the colon with similar but shorter villi in coprodeum. The colon relief is comparable in singing honeyeaters (*Meliphaga virescens*), folds in the coprodeum are, however, irregular and much flatter. (2) A very different pattern was found in colon and coprodeum of galahs and kookaburras (*Dacelo gigas*): their mucosae are mammalian in appearance with crypts opening directly on flat villus-free surfaces. (3) Senegal doves (*Streptopelia senegalensis*) are intermediate with respect to the previous patterns. (4) The mucosa of the colon and coprodeum of the emu are much more elaborate than in the other species, the colon having many closely spaced, tall folds encircling the organ at right angles to its long axis. Each fold contains a core of submucosa and smooth muscle of the muscularis mucosae. Numerous villi extend from the folds producing a pattern like plicae circulares in the mammalian small intestine. The coprodeum is similar, except for less extensive development of the plicae-like folds. Examples of the mucosal patterns are shown in Fig. 6.2.

Concerning the possible relation of structure of the coprodeum and large intestine to function, it would appear that the lower gut of zebra finches, singing honeyeaters, and particularly the emu, contains a relatively larger surface area than

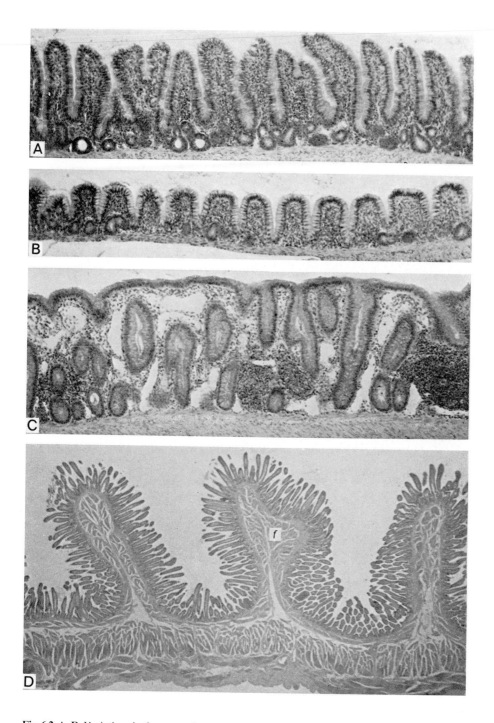

Fig. 6.2. A–D. Variations in the mucosal patterns of coprodeum and colon. **A** Colon of the zebra finch. **B** Coprodeum of the zebra finch. **C** Colon of the galah. **D** Colon of the emu showing tall folds (*f*). A – C × 120, **D** × 10. Reproduced from Johnson and Skadhauge (1975)

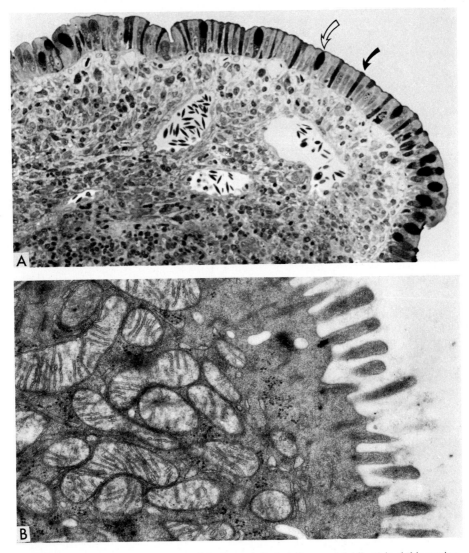

Fig. 6.3. A, B. Mucosa of the coprodeum of the domestic fowl. **A** Shows a toluidin stained thin section. *Open arrow* indicates a goblet cell, *dark arrow* a "dark" cell. **B** Shows an electronmicrograph of the luminal end of a dark cell which is rich in mitochondria. $A \times 600$, $B \times 10000$. Courtesy of Dr. Kjeld Møllgård

that of Senegal doves, galahs, and kookaburras. The relatively large surface area helps in explaining the capacity for resorption of salt and water by the coprodeum and large intestine as measured in the emu (see p. 111).

Recently Eldrup et al. (1979, 1980) have examined by light microscopy of thin sections, electron microscopy and freeze fracture the coprodeum of the domestic fowl exposed to a high and a low NaCl diet Fig 6.3. They observed that Na depletion is associated with an augmentation of the number of dark mitochondria-rich cells of the epithelium Fig. 6.3. In the luminal membrane of these cells Na depletion induced rod-shaped particles. As they span the membrane, the authors suggest they might function as Na-channels. Other studies have shown that Na depletion increases the permeability to Na of the luminal membrane (see p. 118).

6.1.2 Storage of Urine in Coprodeum and Large Intestine

Before discussing the role of the cloaca in resorption from urine, it would seem most important to know which part of the cloaca is actually exposed to urine, and whether there is any dependence upon the diuretic state of the bird. It would also seem worthwhile to compare the osmotic and ionic concentration of ureteral urine with that voided from the cloaca in various types of osmotic stress.

On visual inspection, uric acid can be seen to move retrogradely into the coprodeum and large intestine of the domestic fowl both in the dehydrated and water- and salt-loaded states. Threads of uric acid may even be observed in the caeca close to the junction with the intestine. Evans blue and methylene blue, instilled into the proctodeum, have been observed to move backward even to the caeca (Skadhauge 1968). This was also observed by Browne (1922). Bell and Bird (1966) could find no uric acid, by the murexide reaction, in the caecal contents of the domestic fowl. The reason is probably that they analysed the bulk of caecal material at the blind end, where uric acid never comes.

A special study was carried out in dehydrated domestic fowl (Skadhauge 1968), in which high plasma concentration of inulin was maintained for several hours and the concentration of fractions of the urine–faeces mass (the supernatant after centrifugation) in the cloaca was analysed for inulin. The inulin concentration was observed to be almost unchanged from the lower end of the coprodeum to the lower end of the large intestine; it decreased anterior to this, but was again higher in the caecum. There would thus seem to be ample evidence that the urine moves orally. Urine also moved orally into the coprodeum and large intestine both in the salt- and the water-loaded states. This would indicate that the regurgitation is not regulated by opening or closure of the sphincter between urodeum and coprodeum. Some kind of regulation of the regurgitation does, however, seem to exist since Skadhauge (1968) observed that any fluid or gel instilled in the cloaca during dehydration was retained for hours by the firmly closed anus, whereas the same amount instilled in a normally hydrated bird was quickly expelled. The oral or reverse movement has been confirmed by radiography using X-ray dense material excreted through the kidney in the domestic fowl (Akester et al. 1967). The X-ray dense material also moved into caeca (see p. 35). A quantitative estimate of the fraction going that far was, however, not possible. Reverse movement after urography in the domestic fowl was also observed by Koike and McFarland (1966) and in the roadrunner by Ohmart et al. (1970). Nechay et al. (1969) showed the X-ray dense material to appear in the cloaca of the domestic fowl within 2 min after i.v. injection; in one chicken it took 8 min for the material to reach the anterior end of the large intestine.

The reverse movement influences the composition of the urine–faeces mass. In the dehydrated domestic fowl the average coprodeal osmolality is 563 mOs, compared with 426 mOs in the large intestine. During water loading, coprodeal osmolality is 143 mOs, and 192 mOs in the large intestine. During salt loading, the respective figures are 480 and 466 mOs (Skadhauge 1968). The intestinal wall is thus exposed to rather large osmotic differences in both directions. The composition of the anal and oral ends of droppings defecated by dehydrated and saline-loaded zebra finches is shown in Fig. 5.4. A large fraction of the resorptive area of the lower gut is thus exposed to hyperosmotic material. This is also the case in birds of prey in which the coprodeum functions almost like a bladder containing liquid urine. As measured in the kookaburra, the osmolality is, however, very hypertonic (see Table 5.5).

There is considerable difference in the "dropping pattern" of birds. As this is important for the likelihood that the droppings collected under oil express composition of ureteral urine, the anatomy of the droppings of the Australian birds investigated by Skadhauge (1974a) is presented in detail.

6.1.3 "Anatomy" of the Droppings

The droppings could be grouped primarily according to three patterns: rod-shape, snail-shape, and fluid. Most uric acid was deposited in the anal end (see Fig. 5.4). Rod-shaped droppings are passed by the zebra finch, the singing honeyeater, the red wattle bird (*Anthochaera carunculata*), and the galah. In these droppings one end is covered by uric acid and easily distinguished and separated from the oral end. This category also includes the droppings of the emu which are onion-shaped and covered at one end with uric acid. This shape reflects the shape of the coprodeum in vivo. Droppings falling like a snail with the top downwards are produced by crested pigeons (*Ocyphaps lophotes*) and Senegal doves. The centre contains the ureteral part, which is covered with uric acid, and around it the droppings coming from the oral end are located. It is often possible to uncoil the snail and separate the lower from the upper end. Fluid urine and faecal matter is passed by the kookaburra, in which "pure" urine often fell in separate droppings different from the brown-coloured faecal material.

The storage of urine results in movement of salt and of water across the intestinal epithelia. Yet only in the fairly dehydrated state can the urine flow be expected to be slow enough to permit the possible resorption to exert a quantitative influence on the final cloacal output of salt and water. This is suggested by the fact that the maximal osmolality and electrolyte concentration of cloacally voided urine during dehydration and salt loading are not significantly different from ureteral urine, of domestic fowls, collected through funnels (Skadhauge 1968). During hydration the minimal concentration was unchanged. In the hydrated and salt-loaded states these findings most likely are due to the high rate of flow, but it is surprising that it was also observed in the dehydrated state. This observation was also made in the budgerigar (Krag and Skadhauge 1972). These findings are most probably due to the fact that the spontaneously "voided" urine is located most anally in the digestive tract, in the lower end of the coprodeum, and this urine is largely unaffected by the fate of the fraction of the urine that moves orally. If samples of cloacal urine were taken from the coprodeum in the dehydrated fowl, which showed no inclination to void, the osmolality and ionic concentrations were found lower (Bindslev and Skadhauge 1971b) than those of ureteral urine (Skadhauge and Schmidt-Nielsen 1967a). Similar findings were made by Hughes (1970a) on the gull (*L. glaucescens*). A distinction must thus be made between the situations in which the cloacal contents are removed artificially, and those in which the coprodeal contents are voided spontaneously.

6.1.4 Water Content of Faeces

In order to demonstrate the effect of dehydration upon the water content of the total droppings, this problem will be shortly reviewed. The water content of the urine–faeces mass has been measured in several studies (Table 6.1). It appears that the water content decreases in the dehydrated state. How much of this is due to diminished urine output and how much to increased cloacal absorption, cannot be estimated from water content alone. Since dehydration as a rule decreases motor activity (Krag and Skadhauge 1972) and lowers the food consumption (Kellerup et al. 1965), water conservation is larger than what might be reckoned from the fractional decrease in the droppings. The total dry weight of the faeces must be less. It is important to note that in no case has a water content of less than 50% been observed. Since in uric acid suspensions (semi-dry masses) the "matrix potential" only at a water content of 46% or lower begins to raise above zero (Murrish and

Table 6.1. Faeces water content %

Species	Water ad libi. %	Dehydra- tion %	References
Domestic fowl	78	—	Röseler (1929)
	78	—	Yushok and Baer (1948)
	80	—	Osbaldiston (1969a)
	88	—	Dixon (1958)
Domestic turkey	85	—	Scheiber and Dziuk (1969)
	80	—	Vogel et al. (1965)
Budgerigar *(Melopsittacus undulatus)*	86	68	Cade and Dybas (1962)
Galah *(Cacatua roseicapilla)*	83	66	Skadhauge and Dawson (1980a)
Cowbird *(Molothrus ater)*	90	77	Lustick (1970)
Zebra finch *(Poephila guttata)*	85	55	Lee and Schmidt-Nielsen (1971)
Zebra finch	80	65	Calder (1964)
Black-throated sparrow *(Amphispiza bilineata)*	81	57	Smyth and Bartholomew (1966b)
Stark's lark *(Spizocoryx starki)*	74	52	Willoughby (1968)
Grey-backed finch-lark *(Eremopterix verticalis)*	75	51	
Sage sparrow *(Amphispiza belli nevadensis)*	76	62	Moldenhauer and Wiens (1970)
Common tern *(Sterna hirundo)*	80	—	Hughes (1968)
Vesper sparrow *(Pooecetes gramineus)*	82	63	Ohmart and Smith (1971)
Brewer's sparrow *(Spizella breweri)*	80	59	Ohmart and Smith (1970)

Schmidt-Nielsen 1970), compared with the solute potential, it may be concluded that under natural conditions the chemical potential of water (osmolality or freezing point depression) determined on the supernatant of droppings will also be valid for the uric acid/faeces mixture. The matrix potential or "soil potential" denotes the contribution to chemical potential of water created by capillary force.

6.2 Studies of Cloacal Absorption of Ureteral Urine

Storage in the lower intestine modifies the amount and composition of ureteral urine. There are basically three ways of assessing the quantitative effect of this storage.

The first is separation of ureters and lower intestine and quantitative collection of ureteral urine and faeces and comparison with the normal output. The second is installation in vivo of fluids of chyme- or urine-like composition into the segments of the intestine (coprodeum and colon) with which the urine comes into contact, and recording of absorption of the fluid. The third is measurement in vitro of the transport parameters of the epithelium taken from the coprodeum and colon. Proctodeum and urodeum can be excluded in this context, first because urine is not stored there, second because the epithelium is not a resorptive one (see p. 93). Absorption in the caeca ought to be investigated in detail quantitatively; only preliminary studies have been made (see p. 36).

The first method has two modifications. Urine may either be collected after surgical interference simultaneously with faecal material, or the output of urine may be determined by quantitative collection in acute experiments and compared to the amount of natural droppings passed over a period of time. The difference will— ceteris paribus—give an estimate of cloacally resorbed urine. Both methods are susceptible to both fundamental and practical errors.

Until 1970 only a few papers had dealt with in-vivo absorption in the cloaca, and its epithelium had never been studied in vitro. In the past decade in-vivo and in-vitro studies were published. The majority of the investigations have been on the domestic fowl, but the domestic turkey, the domestic duck, the galah, and the emu have also been used.

6.2.1 Separation of Ureteral Urine and Cloacal Faeces

Several surgical techniques have been devised for either exteriorisation of the ureters or colostomy. Several methods for separate quantitative collection of the excreta on unanaesthetised freely moving birds have been devised (Ariyoshi and Morimoto 1956; Bokori 1961; Colvin et al. 1966; Fussell 1960; Imabayashi et al. 1955; Richardson et al. 1960; Paulson 1969; Rothchild 1947). In many cases the birds showed poor growth rate after the operation, sometimes due to sustained infection. The colon has often been reported to be atonic, this leading to retention of faeces. More seriously for studies of osmoregulation, the urine flow rate has been spuriously high, mainly due to polydipsia. Furthermore, Sturkie and Joiner (1959a, b) have shown by surgical implantation of foreign bodies in the cloaca, that such interference in itself disturbs consumption of food and water. In addition, as shown in one study to be quoted in detail below (Hart and Essex 1942), ureteral exteriorisation led to a Cl loss. This is undoubtedly due to prevention of the normally occuring NaCl absorption in the cloaca. A physiological consequence of polydipsia is an exaggerated estimate of ureteral urine flow rate and hence an overestimation of cloacal absorption rate. The existence of cloacal absorption may, however, also be rejected on false criteria from this type of experiment. The reason is that undisturbed birds with free access to water do not pass a concentrated urine, but excrete a urine with an osmolality close to that of plasma. Therefore, any difference that might be observed only during dehydration will escape detection if the operated birds are never exposed to water restriction. Even if a comparison is made of the output of excreta of normal dehydrated birds, and some with separate collection of ureteral urine (Skadhauge 1968), the outcome must be regarded with some reservation. Periods of ureteral collection must be brief since the collection funnels must be rinsed continuously, and the periods of collection from cloaca must necessarily be long due to the discontinuous discharge of faeces.

6.2.2 Indirect Estimate of Cloacal Absorption of Ureteral Urine by Comparison of Normal Rate of Formation of Droppings and Flow Rate of Ureteral Urine

In early studies (Wiener 1902; Milroy 1904) increased production of urine and higher water intake were noted after the operation to separate the ureters, but precise quantitative measurements were not carried out. This high rate of urine flow, which was observed also during acute experiments with collection of urine, led several authors, by comparison to the total excreta in the undisturbed bird, to assume that a high rate of absorption occurs normally in the cloaca (Sharpe 1912; Rehberg 1926; Korr 1939). Others correctly attributed the high rates of urine flow to the anaesthesia (Davis 1927; Young and Dreyer 1933).

Quantitative collections of urine from cannulae were carried out in the domestic fowl by Hester et al. (1940), who noted a pronounced disturbance diuresis, and changed the experimental procedure for collection of urine from exteriorised ureters. They found a 24 h output ranging from 124 to 61 ml (body weight not stated) in birds with free access to water. The water intake was the same for birds with exteriorised ureters and normal birds ranging from 50 to 250 ml/24 h. An average intake of 163 ± 2.8 (S.E.) ml was observed when two normal and three domestic fowl with exteriorised ureters were followed for 23 days. These investigators could thus reject earlier estimates of very high flow rates of ureteral urine, but could not find positive evidence for cloacal resorption. It should be noted, however, that the operation had "the disadvantage that the urine could run back into the coprodeum" (Hart and Essex 1942).

Skadhauge and Dawson (1980a) measured the 24 h rate of cloacal water output in normally hydrated and in dehydrated galahs by quantitative collections of droppings under oil; further the water content (%) of the ileal contents was estimated when a number of birds were killed for other experimental purposes. The findings allow an indirect estimate of the 24 h ileal water output. The 24 h non-uric acid dry matter was estimated in the excreta, and zero dry matter absorption in colon and coprodeum was assumed. In separate experiments (Skadhauge 1974a) the ureteral flow rate was estimated. Although the values are semi-quantitative, the resulting estimate of cloacal water absorption, about 500 μl/kg·h, was not far from that measured by in-vivo perfusion in which the rate of solute-linked water flow was 442 μl/kg·h (Skadhauge 1974b).

6.2.3 Simultaneous Collection of Ureteral Urine and Faeces

In the investigation by Hart and Essex (1942) more precise collections of ureteral urine and faeces in birds with artificial anuses were made. These authors attempted two most relevant experimental procedures: (1) Feeding a low-salt diet (omitting the "normal" addition of 1% NaCl), and (2) Dehydration.

The first procedure resulted always in a large loss of weight with hypochloraemic haemoconcentration which became pronounced after several weeks. Addition of NaCl to the diet had the following effect: "The body weight increased rapidly; the haematocrit and plasma chloride values returned to normal almost immediately". The normal plasma Cl was 110–120 mequiv/l, and haematocrit 30%–40%. Typical values on the low-NaCl diet were Cl: 92–94 mequiv/l and haematocrit 52%–54%. One fowl after 3 weeks on low NaCl diet had lost weight from 1300 to 800 g, and the plasma Na was reduced from 162 to 136 mequiv/l. Hart and Essex have thus revealed an important role of post-renal NaCl absorption in the cloaca, which is of very great physiological importance during low NaCl intake. They also noted an increased water intake if NaCl was not added to the diet. This is in accordance with the polydipsia that results from NaCl depletion in mammals. Sodium depletion may thus be one of the reasons for the polydipsia that was observed in birds with separated ureters and cloaca.

During dehydration the operated birds lost weight more rapidly than normal birds. The total weight loss over 3 days was 66 ± 5 ml/kg for seven normal birds, and 89 ± 9 ml/kg for four fistulous birds. The difference is highly significant. The average saving is 23 ml/kg or 321 μl/h·kg. This value is in agreement with the transport capacity of the cloaca (see p. 109) and the observations of other workers (Dicker and Haslam 1972). Hart and Essex reported in detail the intake of food and water, and production of urine and faeces of two exteriorised birds. Accepting standard values for water content of food and production of metabolic water, total water loss can be divided among urine, faeces, and evaporation for these normally hydrated birds (Table 6.2). It is apparent that a good third of the water is lost by evaporation, and that the water content of the excreta is 30%–40% of the ureteral urine production. The water intake varied from 70 to 175 ml/kg·day.

Dixon (1958) did not find significantly different concentrations of water in the excreta when two different operations to separate ureters and cloaca were carried out on two groups of three hens. In one experiment the ureters were exteriorised

Table 6.2. Urinary and faecal excretion of water and evaporation in domestic fowl after exteriorisation of the ureters. Units: ml/kg day (Mean ± S.E.)

	Hart and Essex (1942)		Dicker and Haslam (1972)
	Bird No. 1	Bird No. 7	Average of 6 birds
No. of days of study	9	4	4
Volume of urine	55 ± 3	80 ± 7	40 ± 4
Water in faeces	17 ± 2	32 ± 6	54 ± 4
Evaporation	31 ± 4[a]	55 ± 9[a]	66 ± 7
Water content of urine-free faeces	75%	85%	80%

[a] Difference between drinking and urine/faeces loss. Metabolic water and water in food not taken into account

according to Dixon and Wilkinson (1957), in the other cloaca was separated between coprodeum and large intestine, which was then sutured to the skin. This leaves urine in contact with coprodeum and, as pointed out by the author, the possibility of contamination by secretions from the oviduct. Since the birds maintained high intake of food and fluid, and thus presumably received sufficient NaCl, and as dehydration was not attempted, only a small difference between non-operated and operated birds should be expected. The average output in the excreta was 5.78 ml/kg·h. The cloacal resorptive capacity from an isosmotic NaCl solution is less than 5% of this value (see p. 109).

Dicker and Haslam (1966) exteriorised the ureters of Rhode Island red male fowl with a mean body weight of 2.5 kg. The birds received a diet containing 1.25% of salt and vitamins. They had a urine flow from the exteriorised ureters about 40 ml/kg·day, which corresponds to near-isotonic urine. The average water intake was 55 ml/kg·day in the normal birds and it stabilised at values 30 ml higher in the exteriorised birds. Since no pronounced polydipsia was apparent in these birds, the difference, as judged from the urine parameters, may reflect loss of cloacal resorption. Dicker and Haslam also observed a rather variable excretion rate of creatinine, which most likely is caused by a variable glomerular filtration rate.

The same authors (Dicker and Haslam 1972) extended their investigation to a quantitative collection of urine and faeces, and an estimate of evaporation, taking into account the fluid intake, the water content of food, and metabolic water. In a group of six operated birds food intake was constant after the operation and equal to that of six controls which also had constant fluid intake. The only difference after the operation was an average extra water intake of 25 ml/kg·h equal to the increase in urine and faeces water loss of 27.4 ml/kg·h. The loss of water in urine and faeces excretion was slightly higher than those of Hart and Essex (see p. 102), but not more than might be caused by different types of food, varying room temperature, etc. Their estimates of cloacal absorption rate, 25 ml/kg·day, are remarkably similar to that of Hart and Essex. It is higher than the value (13.4 ml/kg·day) computed from their values using cloacal-transport parameters as estimated by in vivo perfusion experiments (see p. 146). Considering all the differences, the agreement is acceptable and a normal cloacal absorption around 20 ml/kg·day seems realistic in the non-dehydrated domestic fowl receiving a normal salt intake.

Dicker and Haslam raised further the interesting argument that if the changing GFR and urine osmolality observed by them result in spells of hypo-osmotic urine, the cloacal resorption of H_2O will proceed along an osmotic gradient when hypo-osmotic fluids are brought into the gut. This will augment resorption to exceed that of the solute-linked water flow. This can only be the case if the periods of hypo-osmotic productions are short. Otherwise fluid will be voided as in water-loaded birds, and the cloacal resorption becomes insignificant (see p. 144).

Scheiber and Dziuk (1969) prepared rectal fistulae in Wrolstad medium-white turkeys of 8–10 weeks of age, in which the rectum was "...severed just anterior to the coprodeum...", thus allowing contact of urine with the coprodeum. The turkeys were fed a grower ration, the NaCl content being fairly high since the excretion of Na and K in non-operated birds averaged 8.3 and 20.2 mequiv/kg·day, respectively. These turkeys, however, became polydipsic and grew at less than half the rate of the non-operated controls; the water intake was trice as great as in unoperated birds (600 as compared to 200 ml/kg·day) and the mean osmolality of the urine was only 63 mOs. The water content of the faeces was the same, 85% in fistulated as in normal birds, whereas the average composition of urine and faeces combined in the fistulated birds was 97% water. This study does not elucidate the role of cloacal water absorption. Scheiber et al. (1969) have studied such operated turkeys for 2 months. Seven out of nine birds suffered from colon impaction and megacolon. In all turkeys, including the healthy ones, the water intake remained two to three times higher than the intake before surgery.

In a study on excretion of Ca and P, Brown and McCracken (1965) measured the excretion of Na and K in faeces and urine in colostomised laying pullets. On a food containing 0.26% of Na and 0.73% of K they observed a total excretion of 14% of Na and 69% of K in the ureteral urine, 86% of Na and 31% of K in the faeces.

Due to the difficulties in the interpretation of the results of these experiments it seemed relevant to attack the problem of cloacal absorption in a more direct manner, i.e. by measurement in vivo and in vitro of the transport parameters of the cloacal epithelia. Since the segment of the gut with which urine comes into contact is known (see p. 97), and since the osmotic and ionic composition of the contents has been measured in the domestic fowl (see p. 98), the rates of absorption and secretion of relevant solutes and water can be calculated and compared with the excretion in ureteral urine. Even such calculations leave one problem unsettled: The contribution of water and solutes coming from ileum to the large intestine and coprodeum. The normal flow rate and quantitative composition of the chyme from the ileum must be measured. This remains to be done in a quantitative way. The cloacal fractional absorption/secretion rates of ureteral urine estimated in Chap. 8 are, therefore, maximal values.

It should be noted that concentrations of ions in ureteral urine apply to the supernatant after centrifugation, but renal excretion rates apply to total amount as determined on urine flushed out with a rinsing flow of water.

The fraction of solutes normally sequestered with the precipitate of uric acid/urates should be determined (see p. 90), and the possible release in the cloaca measured.

6.3 In-vivo Instillation and Perfusion Studies

Available evidence will be presented in the following order: First, early studies in which fluid absorption was demonstrated will be reviewed briefly. Second, recent studies elucidating quantitatively the basic parameters of NaCl and water transport will be outlined. Third, the effects of hydration and dehydration, salt depletion, and loading on transmural transport parameters will be summarised. Fourth, the transport of other ions that are abundant in ureteral urine, i.e. ammonium, potassium, and phosphate ions (see p. 114), will be described.

6.3.1 Early Observations

All of these investigations were instillation experiments. This means that no stirring or circulation was used. Sharpe (1923) instilled hypo- to isosmotic NaCl solutions into the coprodeum and colon of urethane-anaesthetised hens. Samples were taken at 1-2 h intervals for a total of 3 to 5 h. Haemoglobin served as a water marker. Sodium chloride and water was observed in these studies. In five experiments with NaCl concentrations ranging from 137 to 147 mM an average of 868 μl H_2O, 207 μequiv Cl, and 174 μequiv Na were absorbed per hour. Unfortunately, the weight of the birds was only stated for some of the experiments with either 1 or 2 kg. The ion absorption rates are in the range of those reported later (see p. 109).

Weyrauch and Roland (1958 instilled 10 ml of an electrolyte solution into the cloaca of Rhode Island red hens weighing 2.2–3.4 kg. By injection of a barium mixture through the anus and X-ray after the fluid was instilled, it was observed to reach as far orally as to the junction between large intestine and the ileum, and caeca. After approximately 1 h, 10% of Na^{24} was absorbed into the blood in five hens after instillation of 0.9% of NaCl. Disregarding backflux of isotope, volume changes, and ureteral excretion, and assuming a body weight of 2.8 kg, the net Na absorption rate can be calculated with 55 μequiv/kg·h. Since this is lower than the true value, as all the disgarded factors would increase the rate, the agreement with this observation and later quantitative measurements (see p. 109) is good. Slightly lower absorption rates were found for I^{131} from NaI solutions.

Skadhauge (1968) did not instil fluids in the lower intestine, but an unabsorbable water marker, inulin-C^{14}, came with the urine into the cloaca of roosters, after being filtered in the kidney. A fairly constant plasma concentration was maintained over several hours by repeated intramuscular injections of inulin-C^{14}. The inulin concentration and the concentrations of Na, K, and Cl, and the osmolality of the supernatant after spinning of the contents along the length of the lower intestine, was measured in seven dehydrated roosters (Fig. 6.4). If the cloacal contents had a higher inulin concentration than the ureteral urine taken by pipettes just before the slaughter of the birds, water absorption from the cloacal contents would have been demonstrated. In this study water resorption was not directly demonstrated by a higher inulin concentration in the urine–faeces mass in coprodeum than in ureteral urine. But an only slightly lower inulin concentration in the major part of the lower intestine supports the concept that there is considerable

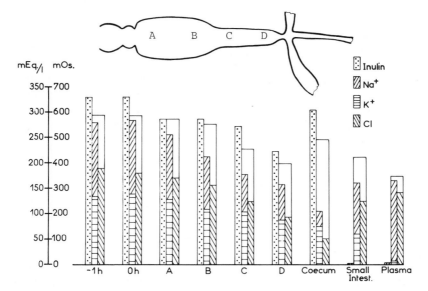

Fig. 6.4. Composition of the contents of coprodeum and colon of the dehydrated domestic fowl. The contents of the coprodeum and colon were divided into four fractions (*A-D*) and analyses made on the supernatant fluids after centrifugation. Ureteral urine was collected 1 h before (-1h) and at the termination (0h) of the experiment. Osmolality of the samples is indicated by the *open lines*. The inulin concentration remained almost unchanged throughout the coprodeum and colon, demonstrating retrograde flow of Urine into these segments of the lower intestine. Inulin concentration relative to the concentration in *A*. Reproduced from Skadhauge (1975)

absorption of water. The flow rate of ureteral urine is approximately 1.2 ml/kg·h in the dehydrated domestic fowl (see Table 5.4). If the flow of faeces from the small intestine is of the same magnitude, this should result in a dilution to 50% of the ureterally excreted urine and inulin if no resorption of water takes place, as the urine is "diluted" with faeces coming from the small intestine. A flow of faeces from ileum of this magnitude in the dehydrated state seems quite reasonable, as Dicker and Haslam (1972) found a faeces excretion of 2.8 g/h with a water content of 80% in normally hydrated fowl when the ureters were separated from the cloaca. The study of Skadhauge (1968) would thus indicate a resorption of approximately 50% of the volume of cloacal urine–faeces. Water and salt are, therefore, absorbed even from the hyperosmotic contents present in the cloaca in the dehydrated state.

Vogel et al. (1965) instilled a Ringer solution (a plasma isosmotic, buffered NaCl solution with glucose) in the coprodeum and colon of five domestic turkeys. Access of ureteral urine and ileal contents was prevented. They measured by direct recollection an average volume absorption of 675 μl/kg·h. Although not reported quantitatively, an absorption of 10%–15% of Na was observed and secretion of K, as the K concentration increased four to five fold. The volume resorption capacity was estimated at approximately 10% of the ureteral urine flow of a normally hydrated turkey.

The absorption of tritiated water (HTO) into plasma from coprodeum of domestic ducks was measured by Peaker et al. (1968) after instillation in 0.7% of NaCl. A maximal number of counts occurred 25–45 min after introduction of HTO, with a loss of 75% of the total instilled activity after 5 min. Since penetration of HTO is most likely limited by mucosal blood flow (Skadhauge 1973), this study demonstrates a good vascularisation of the epithelium.

6.3.2 Basic Measurements of Transport Parameters of NaCl and Water of Coprodeum and Colon

The studies to be reviewed in this section have been carried out by perfusion of coprodeum and colon with either recycling of fluid or in open circuit. The volume and solute transport rates have been calculated from the change in concentration of an unabsorbable water marker (inulin or polyethylene glycol 4000) and the concentration change of the solutes. In the recycling experiments the instilled volume must be measured precisely. In open circuit experiments the inflow rate has to be constant and measured, and the outflow concentration measured after steady state is achieved.

To facilitate comparison with other species and in order to compare values from individuals of the same species, the transport rates have usually been expressed as μl or μequivalents per kg body weight·h. The reference to kg body weight is the best choice, as the same functional segment has been perfused. For other comparisons the simple (serosal) surface area can be used for calculation of transport rates. These perfusion experiments have the virtue of being made under rather natural conditions; important is the presence of normal blood supply. A drawback is that anaesthesia is necessary, and that the geometry is poor for electrical potential measurements.

The following parameters are the most important in characterisation of net salt and water movement across the epithelium when fluids of varying NaCl concentrations and osmolality are brought into the lumen of the intestine: (1) The osmotic permeability coefficients in the two directions. (2) The rate of net Na (Cl) absorption measured as function of the luminal concentration of Na. This parameter obeys saturation kinetics and can thus be described by a maximal flow rate (V_{max}) and a concentration for half maximal flow (K_m). (3) The rate of solute-linked water flow, which is the water absorbed in the absence of an osmotic gradient across the epithelium. This water flow in intestinal epithelia is generally observed to be proportional to the net Na (Cl) absorption rate. It is most likely caused by a local osmotic coupling (Skadhauge 1973; p. 58). Also in the cloaca of the domestic fowl proportionality between net NaCl transport and water absorption was observed (Thomas and Skadhauge 1979a). The rate of solute-linked water flow can therefore be expressed as μl water per milliequivalent of Na or as the Na concentration of the absorbate.

The osmotic permeability coefficient, Pos, has been determined in three studies on the domestic fowl (Skadhauge 1967; Bindslev and Skadhauge 1971a; Thomas and Skadhauge 1979a) and in one on the galah (Skadhauge 1974b). The studies used perfusion with hypo- or hyperosmotic fluids containing impermeable solutes instead of Na. This prevents interference from solute-linked water flow. The results of these experiments are very similar: the Pos was always smaller for water flow from blood to lumen and higher from lumen to blood. In both cases the flow was

Table 6.3. Osmotic permeability coefficient of the coprodeal and large intestinal epithelium. Mean $\pm$ S.E. Unit: $\mu l/kg\,hr\,mOs$

Species	Osmotic agent	Mucosa-serosa[a]	Serosa-mucosa[a]	References
Domestic fowl	Raffinose	3.6	2.3 ± 0.7	Skadhauge (1967)
	Raffinose	5.8 ± 0.5	3.2 ± 0.5	Bindslev and Skadhauge
	$MgSO_4$	4.5	2.5	(1971a)
	Raffinose	—	1.2	Skadhauge and Thomas (1979)
Galah	306 mOs[b]	—	2.2 ± 0.3	Skadhauge (1974b)
(*Cacatua roseicapilla*)	404 mOs[b]	—	1.4 ± 0.2	

[a] Indicates direction of the water flow
[b] Mean osmotic difference

linearly dependent upon the driving force, at least for osmolality differences up to 200 mOs (Table 6.3). In the galah higher average luminal osmolality resulted in lower permeability (see Table 6.3). This phenomenon, designated as rectification, may be caused by a combined effect of shunt pathways, unstirred layers, and serial resistances (Bindslev and Skadhauge 1971a). The determination of Pos is most important as the value permits calculation of the simple osmotic water flow across the epithelium in the two directions when hypo- or hyperosmotic urine flows into the coprodeum-colon segment. The self-diffusional permeability to water was measured by Skadhauge (1967) by the HTO disappearance rate from the lumen. The tritiated water turnover corresponded to 5 ml $H_2O/kg\cdot h$. The permeability coefficient, Pd, measured in cm/s, was 36 times smaller than the Pos expressed in this unit. The possible explanations for the difference have been outlined by Skadhauge (1973). The main reason is that the diffusional permeability is not limited by membrane resistance, but Pd is proportional to, and determined by, the rate of mucosal blood flow.

The net absorption rates of Na and Cl, the electrical potential difference (PD), and the solute-linked water flow in vivo from coprodeum-colon, have been studied in the domestic fowl (Skadhauge 1967; Bindslev and Skadhauge 1971b; Thomas and Skadhauge 1979a; Skadhauge and Thomas 1979), the galah (Skadhauge 1974b), and the emu (Skadhauge et al. 1980). The studies on the domestic fowl (Skadhauge 1967) and on the emu (Skadhauge et al. 1980) were performed with recycling perfusions, the others by open circuit perfusions. When the Na concentration of the luminal perfusion fluid was varied, saturation kinetics of net Na absorption was observed. Thomas and Skadhauge (1979a) observed nearly the same V_{max}: 287 μequiv/kg$\cdot$h in birds receiving a low Na diet as Bindslev and Skadhauge (1971b) found in normally hydrated birds receiving a commercial diet, which contains a fairly high amount of NaCl, 308 μequiv/kg$\cdot$h. The K_m was, however, different: 155 mequiv/l, compared to 99 mequiv/l. The observations of Skadhauge (1967) suggest also saturation kinetics, but with a maximal absorption rate of only 98.5 μequiv/kg$\cdot$h and no precisely defined K_m. In the study of Thomas and Skadhauge (1979a) birds on commercial diet had lower rates of transport of Na and Cl and linear dependence on the luminal Na concentration. The similarity between the saturation

Table 6.4. NaCl absorption rates, and PD, of the coprodeal and large intestinal epithelium.(Transport rates were measured at isosmotic perfusion solutions)

Species	J_{Na} μequiv/kg h	J_{Cl} μequiv/kg h	PD[a] mV	References
Domestic fowl	88 ± 12	52 ± 13	46 ± 5	Skadhauge (1967)
	175 ± 30	42 ± 12	43 ± 5	Bindslev and Skadhauge (1971b)
	31 ± 15	26 ± 14	35 ± 3	Thomas and Skadhauge (1979a)
Galah *(Cacatua roseicapilla)*	88 ± 8	82 ± 11	—	Skadhauge (1974b)
Emu *(Dromaius novaehollandiae)*	541 ± 43	579 ± 40	22 ± 4	Skadhauge et al. (1980)

[a] Lumen negative

kinetics observed on the Na-depleted birds and on birds on commercial diet may be due to the long fasting of 36 h used by Skadhauge (1967) and by Bindslev and Skadhauge (1971b), thus causing the birds to become partly Na-depleted.

The net absorption rates of Na (and Cl), measured in these studies at luminal Na concentrations of 80–100 mequiv/l (isosmotic solutes), are recorded in Table 6.4. The Cl absorption rate was usually parallel with that on Na, but lower. A near electroneutral transport of Na, Cl, and K was observed (Skadhauge 1967). The average transmural electrical PD across the wall of the lower gut, as observed in these studies, is also reported in Table 6.4. The lumen was always negative and Na ion was moved against an electrochemical potential difference suggesting active transport. This was confirmed by in-vitro studies (see p. 117). The Na absorption rate was not only governed by the NaCl balance of the animal (see p. 112) and the luminal Na concentration, but high concentrations of either K or NH_4 exerted an inhibitory action. A K concentration of 80 mequiv/l halved the net Na absorption rate (Skadhauge 1967). A similar reduction was seen after a doubling of the NH_4 concentration of the perfusion fluid from 52 to 104 mequiv/l at a Na concentration of 150 mequiv/l, but only in birds on a commercial diet or on a higher NaCl load (Thomas and Skadhauge 1979a).

In the absence of a difference in osmolality across the gut wall, net NaCl absorption was always followed by a water flow. From isosmotic perfusion solutions a rate of 151 μl/kg·h was observed by Skadhauge (1967), 199 μl/kg·h by Bindslev and Skadhauge (1971b), and 111 μl/kg·h by Thomas and Skadhauge (1979a). These rates for water flow, compared to the simultaneous Na absorption, lead to a hyperosmotic absorbate, and consequently dilute the perfusate. In recycling perfusion experiments even strongly hyperosmotic solutions eventually become hypo-osmotic to plasma (Skadhauge 1967). This allows solute-linked water flow and simple osmotic water absorption to work together. It is highly likely that under normal function of the coprodeum-colon such a dilution takes place in the thin layer of urine around the central faeces core. Very slow collections of perfusate at the wall in the middle of the coprodeum-colon during perfusion at a rate of 9 ml/h of a fluid of 500 mOs have given strong evidence for such local dilution:

Bindslev and Skadhauge (1971b) observed a maximal fall in osmolality and volume marker concentration to 74% and 78% when the recovered bulk fluid only changed to 86% and 91%, respectively.

These findings gave stimulus to experiments in which fluid of an osmolality and concentration of strong electrolytes identical to that of ureteral urine of dehydrated domestic fowl were infused in the coprodeum-colon of hens at the rate of production of ureteral urine (Bindslev and Skadhauge 1971b). A teflon insert simulated the central faeces core. In these experiments a much larger water absorption was observed than when a higher flow rate was used. In rapid flow rate experiments the perfusion fluid would need to be 65 mOs higher than plasma in order to result in zero net water flow, but in the slow perfusion experiments in normally hydrated fowls the value was 153 mOs. This illustrates the importance of total amount of salt and water delivered to the coprodeum-colon as compared to the transport capacity of the epithelium (see Chap. 8).

Besides the studies on the domestic fowl, in-vivo perfusions have been made on the xerophilic parrot, the galah, and on the large flightless ratite bird, the emu. These two species were chosen for their high (galah) and low (emu) renal concentrating ability, respectively (see Table 5.5). It would seem of particular interest to disclose in which way the cloacal transport parameters are adjusted to the osmotic and ionic composition of the urine in these two species. The galah has an average osmotic urine to plasma ratio of 2.7 in the dehydrated state, the emu only 1.4. How will the luminal concentration of ions influence Na transport, and how does the solute-linked water flow correspond to the osmolality and flow rate of the ureteral urine to which their coprodeum-colon is exposed? Furthermore, as both species are abundant in deserts in which the Na content of available food is low, their Na conservation of coprodeum-colon might show a particular adaptation. Finally, another study (Krag and Skadhauge 1972) in a small parrot, the bugerigar, seemed to indicate that this species, in order to survive in the dehydrated state, could not allow itself to lose water as a result of the inflow of the strongly hyperosmotic urine in the cloaca (see p. 147). Measurement of the cloacal transport parameters would be highly interesting. The small size of budgerigar (30 g), however, made cloacal perfusion difficult. The 300 g galah was more suitable for such experiments.

The absorption rates of Na and Cl in the galah coprodeum-colon increased in parallel as functions of the luminal concentrations (Skadhauge 1974b). The net absorption rates were from isosmotic NaCl solutions in the range of those observed in the domestic fowl: Na, 88 ± 9 μequiv/kg·h, and Cl, 82 ± 11 μequiv/kg · h. When higher and lower concentrations were used, Na showed saturation kinetics with a V_{max} of 217 μequiv/kg·h and a K_m of 181 mequiv/l. Cl did not reach saturation. K was secreted into the lumen in these experiments. The transport rate of Cl was slightly reduced when a high (150 mequiv/l) K concentration was used by replacing NaCl with KCl. The transmural water absorption from isosmotic perfusion solutions was higher than in the domestic fowl: 442 ± 31 μl/kg·h from NaCl perfusion solutions. The water flow was still high: 337 μl/kg·h when NaCl was replaced by KCl. In the latter experiments both Na and K were absorbed, and the μl H_2O/mequiv Na + K + Cl absorbed/secreted was identical to the value observed with NaCl perfusion solutions.

110

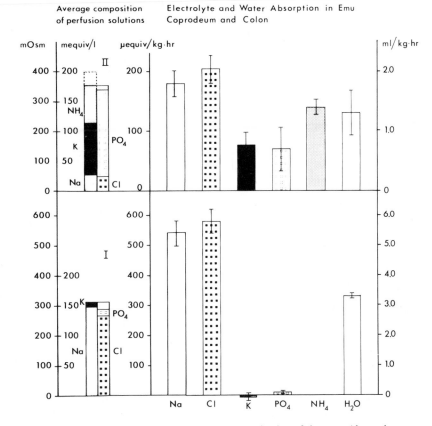

Fig. 6.5. Absorption of electrolytes and water by coprodeum and colon of the emu. Absorption was measured from two perfusion fluids. *To the left* is indicated the osmotic and ionic composition of the perfusion fluids. *I* is a Krebs-phosphate buffer, *II* has a composition similar to ureteral urine of dehydrated domestic fowl (Skadhauge 1977). The rates of absorption of ions and water from the two perfusion fluids are indicated *on the right*. Reproduced from Skadhauge et al. (1980)

The transport parameters of the galah cloaca seem well adjusted to receive hyperosmotic fluids. The Pos was small (see p. 108), and the solute-linked water flow per absorbed Na ion high. A rough quantitative estimate suggests that the galah cloaca can absorb the same fraction of salt and water from urine of dehydrated birds as the domestic fowl (see Chap. 8), about 70% of NaCl and a few per cent of the water.

In the investigation on the emu (Skadhauge et al. 1980), two perfusion solutions were used. Solution I had a near plasma-like composition, as it was identical to the bathing solution of the in-vitro experiments (see p. 116). Solution II had a simulated composition of urine from dehydrated birds (see p. 70), i.e. a slightly hyperosmotic fluid of low concentrations of Na and Cl with K, NH_4 and PO_4 as the major ions. The average electrolyte and water transport from these two solutions of four dehydrated emu chicks (body weight ca. 17 kg) is depicted in Fig. 6.5. The isosmotic

NaCl solution (I) resulted in NaCl absorption rates considerably higher than in the other species: Na, 541 ± 43 μequiv/kg·h, and Cl, 579 ± 40 μequiv/kg·h. The water absorption was $3.36 \pm$ ml/kg·h, which is so large that the absorbate was near-isotonic. From the urine-like perfusion solution (II) the absorption rates of Na and Cl were close to 200 μequiv/kg·h. The NaCl concentrations of the absorbate was still near-isosmotic resulting in a water absorption of 1.30 ± 0.38 ml/kg·h. An important absorption of K, NH_4, and PO_4 also occurred. The K_m and V_{max} for Na absorption was estimated from a Hofstee-plot of the data. An apparent K_m for Na of 25 mequiv/l was found from the data of solution (II). The findings indicate that in the emu the renal and cloacal functions are also matched to benefit maximal salt and water conservation during dehydration and/or salt lack (see Chap. 8). Diuresis induced by the anaesthesia prevented unfortunately a good estimate of normal urine flow rates in these birds, so precise evaluation of the cloacal salt and water transport capacity as compared to the renal output cannot be made.

6.3.3 Changes of Parameters of Cloacal Transport Induced by Hydration and Dehydration and by Salt Loading and Depletion

Comparisons between the cloacal transport parameters of the normally hydrated birds and birds dehydrated for 48 h to an average weight loss of 7% were carried out systematically on the domestic fowl (Bindslev and Skadhauge 1971a, b), and a few measurements were made on the galah (Skadhauge 1974b). The Pos in the serosa-mucosa direction, the maximal Na transport rate, the PD, and the reflection coefficient to Na remained unaffected by the level of hydration, whereas the affinity (the K_m) for net Na absorption, and the Na concentration of the solute-linked water flow changed. Taken together, the changes in these values significantly affected net cloacal water absorption as estimated by computer calculations. The fractional water transport from a simulated flow of ureteral urine of dehydrated fowls was changed from a loss of 14% of the water to a gain of 6% (see Chap. 8). The effect of dehydration was also directly measurable by slow perfusion experiments (Bindslev and Skadhauge 1971b). The rate of solute-linked water flow increased by 38%, and the hypertonicity of incoming fluid, which resulted in zero net water transport, was augmented by 22%. No conspicuous effects of dehydration were disclosed by the experiments with the galah (Skadhauge 1974b).

The effects of different stable levels of NaCl intake have been investigated in the domestic fowl (Thomas and Skadhauge 1979a). The birds received either wheat and barley (low NaCl diet), commercial chicken food (medium NaCl diet) or commercial chicken food supplemented by an oral load of 10 ml 9% NaCl/kg body weight for 2 days. A similar salt load was used by Skadhauge (1967).

The net transport rates for Na were measured in the study of Thomas and Skadhauge (1979a) from perfusion solutions which contained four different concentrations of Na: 0, 30, 75, and 150 mequiv/l. The high Na concentration was also used together with a high NH_4 concentration. Cloacal Na absorption of birds

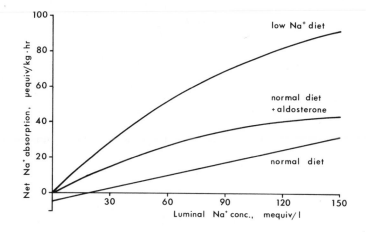

Fig. 6.6. Rate of sodium absorption as a function of sodium concentration in the coprodeum and colon of the domestic fowl. A low NaCl diet augments the rate of sodium absorption. Chronic aldosterone injections only partly restore sodium absorption to the values attained during sodium depletion. Reproduced from Thomas and Skadhauge (1979a, b)

in the three dietary states are depicted in Fig. 6.6 as functions of steady-state luminal concentration of Na. It is seen that the Na absorption rate was much lower for the medium and high NaCl diets, and linearly related to the luminal Na concentration, whereas the higher Na transport rates of the low NaCl diet birds followed saturation kinetics. As the Cl absorption rate followed that of Na, the net NaCl absorption rate due to dietary changes was a function of the NaCl concentration. Larger changes of NaCl absorption occurred from medium to low NaCl diet than from high to medium diet. During perfusion with the solution of high Na concentration the increase in Na + Cl absorption rate was 22 μequiv/kg·h from high to medium NaCl diet, but 81 μequiv/kg·h from medium to low NaCl diet. The main effects of the NaCl content of the diet thus proceeds in the range from low to medium NaCl diet. A similar but even more pronounced pattern was seen when the same diets were used on birds for in-vitro experiments in which coprodeal Na transport was measured in the Ussing chamber (see p. 117). In the experiments of Skadhauge (1967) NaCl loading reduced net Na transport by 68 μequiv/kg·h and net Cl transport by 53 μequiv/kg·h, i.e. by slightly higher values.

The main conclusion of these experiments is that NaCl depletion and/or loading play an important regulatory role in absorption of Na and Cl in the coprodeum-colon of the fowl, and indirectly also determine net transport of water, K, and NH_4, as these fluxes are linked to the Na absorption rate (see p. 115). The increase in net Na and Cl absorption from high to low NaCl intake centers around 50 μequiv/kg·h for perfusion solutions of medium and high Na concentrations. The increase measured from luminal solutions of low Na concentration, characteristic for urine of NaCl depleted birds, may, although small, 11–13 μequiv/kg·h, be more important for the NaCl conservation of the animal, as this capacity is large compared with the renal excretion rates.

6.3.4 Absorption/Secretion of K, NH₄, and PO₄

As indicated in Chap. 5, these ions form the majority of osmolality of ureteral urine, at least of dehydrated, granivorous birds on a low NaCl intake. Three questions are therefore relevant: (1) What are the absorption and secretion rates of these ions in relation to their own concentrations and to the transport of the primarily actively transported solute, the Na ion? How does the cloacal absorption capacity compare with the rate of ureteral output for these ions? (2) How do these ions affect the absorption of Na and of water beyond the contribution to osmolality? (3) Does their presence as abundant ions lead to a reflection coefficient of the cloacal wall lower than the value of unity observed with inert osmotic constituents such as raffinose?

A lower reflection coefficient might apply if the gut wall were highly permeable to these ions or became more permeable due to their presence. In other epithelia the paracellular shunt pathway has been opened both as a result of hyperosmolality per se and by presence of specific solutes.

These questions have been elucidated for the domestic fowl in the study of Skadhauge and Thomas (1979), and for the emu by Skadhauge et al. (1980). When coprodeum-colon of the domestic fowl was subjected to fairly high concentrations of NH₄ (52 and 104 mequiv/l), K (55 mequiv/l), and PO₄ (100 mM), net secretion of K and net absorption of NH₄ and PO₄ always occurred (Skadhauge and Thomas 1979). The K secretion and the NH₄ absorption rates were linearly correlated with the net Na absorption rate, as was Cl (Fig. 6.7). The rate of PO₄ absorption was lower, ranging from 0.7 to 8.4 $\mu M/kg \cdot h$ for all perfusions and dietary states, being thus unrelated to these parameters.

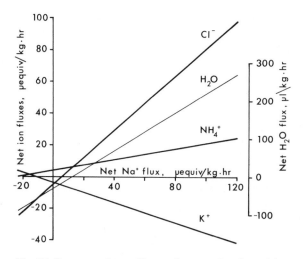

Fig. 6.7. Transport of several ions and water as functions of the rate of sodium absorption in coprodeum and colon of the domestic fowl. The rate of absorption (secretion) of Cl, NH₄, K, and water can be described as linear functions of the net sodium absorption rate. The total transport of the measured ions (including HPO₄, which underwent little net transport) is electroneutral. The water absorption is measured at a luminal osmolality equal to plasma osmolality. Reproduced from Skadhauge and Thomas (1979)

For a "middle" Na absorption rate of 50 μequiv/kg·h, as occurring from an isosmotic (medium Na) perfusion fluid (C_{Na} = 75 mequiv/l), the NH_4 absorption rate was 13 μequiv/kg·h, the K secretion rate 20 μequiv/kg·h, and the Cl absorption rate 35 μequiv/kg·h (rounded values). Doubling of the NH_4 concentration did increase the NH_4 absorption rate by 70% on the average. The average water absorption was 111 μl/kg·h.

In the emu (Skadhauge et al. 1980) virtually no K or PO_4 were transported from the Krebs-phosphate perfusion fluid. From a perfusion medium originally containing K, 84 mequiv/l; NH_4 100 mequiv/l; and PO_4, 136 mM, the following ion absorption rates were observed: K, 76 μequiv/kg·h; NH_4, 140 μequiv/kg·h, and phosphate, 71 μequiv/kg·h. The concentration of these ions remained fairly constant throughout the perfusion, whereas the concentrations of Na and Cl fell. It may therefore be concluded that the osmolality of the perfusion fluid would have risen were these ions not absorbed, and water absorption would have been impaired. The absorption of these ions in the emu coprodeum-colon is therefore contributing to the total "solute-linked" water flow.

Fractional absorption of these ions from ureteral urine in the coprodeum-colon of the domestic fowl will be discussed in Chap. 8.

6.4 In-vitro Studies of Transport Parameters of Coprodeum and Colon of the Domestic Fowl

Important knowledge has been gained from the in-vivo perfusion studies reviewed in the previous sections, particularly with respect to physiological transport parameters. The absorption rates of ions and water across the epithelium of the whole organ have been measured as functions of dietary states of the bird and composition of the luminal contents. For better elucidation of basic parameters of ion transport by the epithelia and determination of unidirectional ion fluxes the in-vivo studies must be supplemented by in-vitro methods. In these the unstirred layers can be reduced and the transepithelial driving forces controlled. Experiments in which the epithelium of either coprodeum or colon was mounted in the Ussing chamber (Ussing and Zerahn 1951) will be outlined in the following section. In this technique sheets of the isolated epithelium, usually around 1 cm², are mounted between two half-chambers containing an oxygenated, thermostated buffer of NaCl solution of plasma-like composition, for example a Krebs-Ringer medium. The electrical potential difference (PD) created by the ion transport of the tissue can be short-circuited and the current recorded, usually by an automatic device. This unclamps the tissue at regular intervals for determination of the PD. After correction for bath resistance the true short-circuit current (SCC) of the tissue and the tissue resistance can be calculated. In this chamber unidirectional fluxes of isotopes in the two directions can be measured, generally between fluids of identical composition and during short-circuiting. Any deviation of the isotope flux ratio (Ussing 1949) from unity will indicate a mediated or active transport as there is no external driving force. The term active transport describes a direct coupling between flow of the substance in question and expenditure of cellular metabolic energy (ATP). For several epithelia the SCC was equal to the net Na transport rate as first observed on the frog skin by Ussing and Zerahn (1951). Measurement of the SCC is thus important in the study of epithelial transport.

Two additional in-vitro techniques will be referred to in this section. The first is the inflow technique described by Schultz et al. (1967). Here the mucosal surface of the epithelium is exposed to a solution containing the isotope of the ion to be measured and an extracellular volume marker, for example polyethylene glycol-^{14}C. After 0.5–0.7 min, when the uptake is still linear as a function of time, the radioactive solution is washed off with a cold solution without isotopes. The radioactivity of both isotopes remaining in the tissue is measured. The cellular uptake is calculated after subtraction of the extracellular activity, as corrected for by the water marker. Obviously, the technique depends upon absence of extracellular trapping of the isotope of the transported species and an equal wash-out of the two isotopes. The other technique is time resolution of the PD change, when abrupt changes in composition of the bathing medium on the mucosal side are forced upon an epithelium in which the active generation of PD is stopped by a metabolic inhibitor (ouabain on the serosal side). This diffusion-potential method permits measurement of permeability to single ions across the mucosal membrane (see Bindslev 1979), as the PD depends upon diffusion across the luminal cell membrane.

The combined utilisation of these methods not only permits measurement of actual rates of active and passive transport parameters for ions, but experiments with different concentrations of ions and voltage clamping experiments allow estimates of ion permeability coefficients. Furthermore, mechanisms of changes induced by different functional states of the epithelium, or by action of hormones, can be investigated.

Isolated sheets of coprodeum epithelium can be prepared under a stereo-microscope dissecting the gut wall at the layer of lamina muscularis mucosae. This has been done in the domestic fowl (Choshniak et al. 1977; Lyngdorf-Henriksen et al. 1978; Bindslev 1979) and in the galah (Skadhauge and Dawson 1980b). The colonic epithelium can be isolated in the domestic fowl by scraping with a razor blade (Lind et al. 1980a).

The basic characteristics of the in-vitro preparation of coprodeum of the domestic fowl was studied by Choshniak et al. (1977), who used white Plymouth Rock laying hens. Two levels of Na intakes were used, either wheat and barley—the low NaCl diet, or commercial chicken food—the high NaCl diet. Both diets were for some experiments supplemented with an oral NaCl load. The average Na intake was for these four diets calculated for a 3 kg bird (mequiv/bird·24 h): wheat or barley 0.4, commercial food 18, wheat and barley NaCl 46, commercial food + NaCl 64 mequiv/bird·24 h. The electrical transport parameters, SCC and PD, were high on the low Na diet: 284 ± 11 μA/cm^2 and 37.3 ± 2.2 mV. These values were strongly decreased by the commercial diet + NaCl to 19.0 ± 1.6 μA/cm^2 and 3.1 ± 0.4 mV. The commercial diet and wheat and barley + NaCl resulted in the same low values. Furthermore, as a function of time SCC and PD fell to zero within 1–2 h in birds on high NaCl diet, whereas these values in those on low Na diet remained high for 3–5 h, provided that glucose was present in the medium (Fig.

116

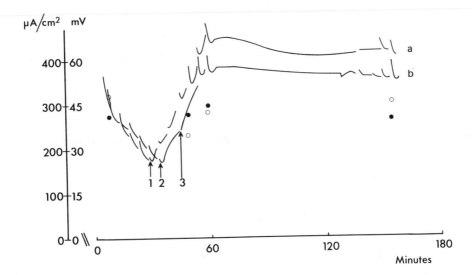

Fig. 6.8. Time course of electric potential difference (PD) and short-circuit current (SCC) of the isolated mucosa of coprodeum from the domestic fowl on a low NaCl diet. Two preparations *a* and *b* were incubated in Krebs-phosphate buffer without glucose. At *arrow 1* glucose was added to 15 mMol on the mucosal side of one preparation *a*. At *arrow 2* glucose was added to 15 mMol to the serosal side of the other preparation *b*, and at *arrow 3* glucose was added to its mucosal solution to 15 mMol. ● PD values belonging to *a*; ○ PD values belonging to *b*. Reproduced from Choshniak et al (1977)

6.8). The majority of the glucose effect seemed to be due to the provision of metabolic energy, since the effect was present after addition of glucose to the serosal side alone, and galactose could not replace the effect of glucose when added to the mucosal side. Unidirectional net fluxes of Na and Cl were measured in short-circuited states in birds on diets both low and high in NaCl. In both cases the flux ratio for Cl did not deviate from unity, whereas the net Na transport was close to the SCC (Fig. 6.9). The SCC is thus predominantly carried by an active transepithelial transport of the Na ion. The Na inflow across the luminal membrane did not deviate significantly from that of the Na isotope flux from mucosa to serosa, suggesting that nearly all Na ions gaining access to the cells are extruded on the serosal side. The back-flux of Na from serosa to mucosa proceeded most likely through a paracellular shunt. In the paper of Choshniak et al. (1977) the back-flow was observed to be linear as a function of the electrical driving force (by voltage clamping); in experiments in which several Na concentrations were used (Lyngdorf-Henriksen et al. 1978) the back-flux was also a linear function of the Na concentration. The apparent permeability was the same in the two experiments: 2.9×10^{-6} cm/s.

The inflow of Na across the luminal membrane was only partially reduced on high Na diet in the experiments of Choshniak et al. (1977). In later work by Bindslev (1979) the range of inflow was high in birds on low Na diet, 5–22 μequiv/cm²·h, and low 0–0.8 μequiv/cm²·h, in high Na-diet birds. The transepithelial mucosa to serosa flux of Na was 13 μequiv/cm²·h (Fig. 6.9). The difference between the two studies is

117

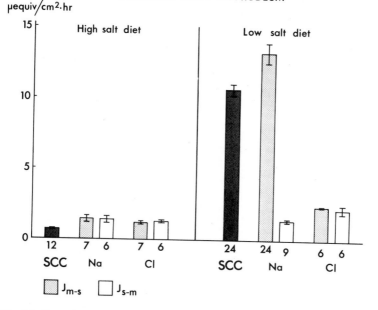

Fig. 6.9. Short-circuit current (SCC) and unidirectional fluxes of Na and Cl across the isolated coprodeal mucosa of the domestic fowl. The preparation is from a bird on a low NaCl diet and the SCC is near or equal to the net absorption of Na. The flux ratio of Cl is unity. A high NaCl diet suppresses net Na transport and SCC is reduced to a value close to zero. Based on data from Choshniak et al (1977)

at present unexplained, but is believed to be caused by lower food intake in some of the birds of the earlier study. This resulted most likely in a partial NaCl depletion.

Two other aspects of transport function of this epithelium have been outlined: the effect of inhibitors (Lyngdorf-Henriksen et al. 1978; Bindslev 1979), and the concentration dependence of unidirectional Na fluxes in the mucosa to serosa direction. Both Na transport and SCC were suppressed by addition of ouabain 10^{-4} mM/l to the serosal side, or amiloride 10^{-4} mM/l to the mucosal side. Ouabain acts by inhibiting the Na/K-activated ATPase, and amiloride by blocking Na-channels in the luminal membrane. Amiloride often resulted in a reversal of the current. This may be due to a small K secretion. The unidirectional Na transport from mucosa to serosa followed saturation kinetics with a V_{max} of 13.4 μequiv/cm²·h and a K_m of 6.3 mequiv/l in coprodeum from birds on low Na diet. The K_m for mucosa-to-cell transport of Na in the inflow experiments was similar, 5.1 mequiv/l (Bindslev 1979). Mucosal dilution experiments confirmed a high permeability to Na of the mucosal membrane in birds on low Na diet (Bindslev 1979). This effect was reversibly blocked by amiloride. On high Na diet the permeability was much lower, with a higher K_m 51 mequiv/l.

The conclusion from these studies is that ion transport across coprodeum is dominated by active absorption of Na that follows saturation kinetics. It is limited by the permeability of the luminal membrane which has a saturable inflow. The suppression of the Na transport by salt loading seems largely to proceed by a

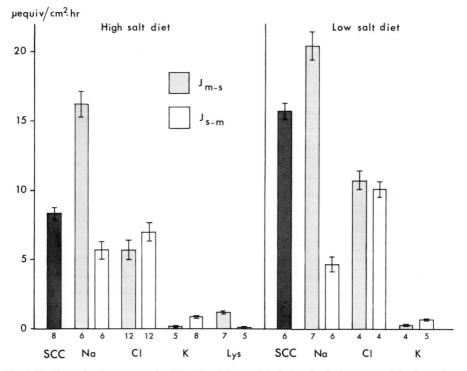

Fig. 6.10. Short-circuit current and unidirectional fluxes of the isolated colonic mucosa of the domestic fowl. Fluxes of Na, Cl, K, and lysine (*Lys*) are shown. The SCC for birds on a low NaCl diet is high and near equal to the net Na absorption rate. The flux ratio for Cl is unity; there is a small secretion of K. In preparations from birds on high NaCl diet the SCC is still fairly high, but only in the presence of both glucose (15 mMol) and leucine (4 mMol) and lysine (4 mMol) in the bathing media. The SCC is equal to the net Na absorption rate; there is a small secretion of both K and Cl, and absorption of lysine. Reproduced from Lind et al. (1980a)

reduction of this permeability. Some other ions, at least Cl, are absorbed by the driving force created by the PD resulting from the net Na transport. The Cl flow proceeds largely through a paracellular shunt.

The electrical parameters and the ion fluxes have been investigated in the colon of the female domestic fowl by Lind et al. (1980a; see Fig. 6.10). In birds on diets low in NaCl the behaviour of the colon is remarkably similar to that of the coprodeum with an average SCC of 360 μA/cm^2 corresponding to 13.5 ± 0.6 μequiv/cm^2·h. This current is largely carried by the net Na transport (see Fig. 6.10). The average PD is lower than in coprodeum, around 21 mV, reflecting a smaller resistance. As in the coprodeum the SCC was suppressed by amiloride and it is metabolically dependent on the presence of glucose. Galactose, lysine, or leucine did not affect the current. However, on diets high in NaCl the ion transport had a pattern very different from the coprodeum (see Fig. 6.8). The basic Na transport and the SCC were low but became high, around 200 μA/cm^2, after full stimulation with galactose, leucine, and lysine. Such sugar and amino acid stimulation of Na absorption is well known from

119

the small intestine of mammals (Crane 1977). The SCC was still accounted for mainly by the net Na absorption rate. The current was insensitive to amiloride.

Thus, in the colon, NaCl loading seems to wipe out a Na transport caused by a high permeability of the mucosal membrane. Instead, there is induced a new type of Na transport that interacts with sugar and amino acids. At the same time sugars and amino acids undergo a limited net absorption (Lind et al. 1980b).

6.4.1 In-vitro Studies in the Galah

In order to test whether the xerophilic, granivorous parrot, the galah, has a special adaptation, the electrical parameters and isotope fluxes of Na and Cl of the coprodeum and colon were investigated in the Ussing-chamber (Skadhauge and Dawson 1980b). The parameters were studied as functions of the salt and water intake of the animals. The electrical parameters were further elucidated as functions of ion concentrations of the bathing media. Finally, the effect of drugs was investigated. The main conclusion from this investigation is that coprodeum and colon of the galah function remarkably similarly to those of the domestic fowl. The SCC was carried by the net Na absorption both in coprodeum and colon, and the flux ratio for Cl was unity (see Table 6.5). The net rate of absorption of Na from the coprodeum of birds receiving mixed seeds (low NaCl diet) was lower than in the fowl, largely due to a higher back-flux of Na. The PD was less than half of that of the domestic fowl due to a 40% lower electrical resistance. The epithelium was thus more permeable, as also revealed by a correspondingly higher serosa-to-mucosa permeability to both Na and Cl. The transport parameters, rate of absorption of Na and PD, of the colon from birds on a low NaCl diet were approximately two-thirds of the values of coprodeum. NaCl loading induced by offering 1% NaCl as drinking fluid suppressed both SCC and net transport of Na by the coprodeum (see Table 6.5), but augmented the colonic-transport parameters. This augmentation may be the result of a better survival in vitro of the tissue of salt-loaded birds. The reaction to amino acids and amiloride was similar to, but not identical with that of the domestic fowl. The colonic SCC was increased by only 18% by addition of leucine and lysine, and was reduced by only 17% by amiloride. The SCC of the coprodeum was completely suppressed by amiloride. Lowering the Na concentration of the medium showed the same K_m as in the domestic fowl, i.e. 5.7 mequiv/l. The

Table 6.5. Comparison of transport parameters between the Galah *(Cacatua roseicapilla)* and the domestic fowl. All values from coprodeum from birds on a low NaCl-diet. From Choshniak et al. (1977) and Skadhauge and Dawson (1980b)

	Galah	Fowl	Unit
Sodium transport in mucosa-serosa direction	7.2 ± 0.7	13.2 ± 0.7	μequiv/cm^2 h
Short circuit current	210 ± 15	284 ± 11	μA/cm^2
Electrical potential difference	19 ± 2	38 ± 2	mV
Resistance	90	131	Ohm cm^2
Permeability to Na	4.5	2.7	10^{-6} cm/s
Permeability to Cl	6.7	4.4	10^{-6} cm/s

coprodeum was resistant to replacement of NaCl with KCl. A high K concentration (73 mequiv/l) reduced SCC by 15% slightly higher than in the domestic fowl (K. Holtug and E. Skadhauge, unpublished). High concentration of either NH_4 or PO_4 reduced the resistance—but induced no consistent change in SCC.

The intestinal transport parameters thus function largely as in the domestic fowl. The lower gut of both these species seems to function effectively in conservation of NaCl doubtless an adaptation to a granivorous diet of low Na content. The limited interference of high luminal concentrations of other ions on net absorption of Na is advantageous considering the high renal-concentrating ability of the galah (see Table 5.5).

6.4.2 In-vitro Studies in Ducks

Because of the obvious interest, the effects of salt loading on transport of Na by the lower intestine in species with salt glands, both domestic ducks and wild mallards, which tolerate high intake of NaCl were either NaCl-loaded or fed wheat and barley (B.G. Munck and E. Skadhauge, unpublished experiments), and coprodea were mounted in the Ussing chamber. The epithelium of the coprodeum of these ducks is far more difficult to dissect than that of the domestic fowl; it is tough, and tissue resistance is higher. Stable values of SCC and PD were measured at only 20–30 $\mu A/cm^2$. As in the fowl, the SCC remained high in the colon, both on a low and high NaCl intake. Major electrolyte resorption will therefore take place in this organ.

6.4.3 Comparison of In-vivo and In-vitro Experiments

When in-vivo and in-vitro studies are viewed together, some discrepancies have an obvious solution, others not. The net rate of absorption of Na was reduced to zero in the coprodeum of salt-loaded birds in vitro, but was reduced only by 50% in the combined coprodeum–colon segment in vivo. This is a natural consequence of maintenance of Na transport in colon, which represents around 50% of the total surface area. The difference in K_m, low in vitro, high in vivo, cannot possibly be accounted for by a difference in unstirred layer thickness. A thick unstirred layer can lead to a spuriously high estimate of K_m. Even if the unstirred layer is assumed 2 mm thick in the in-vivo experiments, the K_m would not be brought near to in-vitro value (Thomas and Skadhauge 1979a). A part of the explanation is a high K_m in the colon of domestic fowls on a high NaCl intake (K. Holtug and E. Skadhauge, unpublished).

6.5 Effects of Aldosterone on Cloacal Transport

The effect of NaCl balance on ion transport by the cloaca, as described in previous sections, is large. It is likely to be mediated at least partially by endocrine regulation.

The dominant adrenocorticoid hormones in birds are aldosterone and corticosterone (see Chap. 9). In this section the effects of aldosterone on Na transport and other parameters of the epithelium of the colon and coprodeum are reviewed. Cloacal perfusions have been carried out in vivo after acute (Thomas et al. 1975, 1979) and after chronic (Thomas and Skadhauge 1979b) injections of aldosterone. In-vitro responses of both the colon and coprodeum to acute injections of aldosterone have been studied (Thomas et al. 1979). The in-vitro results were obtained after injection of the hormone in vivo. Addition of aldosterone to the incubation medium during the experiments did not cause significant effects (Thomas 1973; Thomas et al. 1979; Skadhauge and Dawson 1980b), undoubtedly because in-vitro preparations, at least partially, barely function for the time, about 4–5 h, needed for full augmentation of Na transport by aldosterone. In acute in-vivo experiments a single aldosterone dose of 120 $\mu g/kg$ was injected intramuscularly by Thomas et al. (1975, 1979). This dose is sufficient to cause a maximal effect for several hours. Thomas et al. (1979) used white Plymouth Rock laying hens, which received either a medium NaCl diet (commercial food) or this diet supplemented by an oral load of 10 ml 9% NaCl per kg body weight for 2 days (high NaCl diet). Commercial food resulted in an intake of 12 $\mu equiv$ Na/100 g food. The effect on Na transport of the aldosterone injections is shown in Fig. 6.9. Over a period of 5–6 h the net Na absorption rate increased fivefold in birds on the medium NaCl diet and tenfold in those on high NaCl (see Fig. 6.11) with mean values of 12 ± 2 $\mu equiv/kg \cdot h$ and 3 ± 1 $\mu equiv/kg \cdot h$ in medium and high Na diet birds, respectively, before aldosterone injection, and 52 ± 7 $\mu equiv/kg \cdot h$, and 30 ± 6 $\mu equiv/kg \cdot h$ at the "plateau", 4–5 h after the injections. The K transport was augmented from a near zero net absorption to secretion of 11 ± 2 $\mu equiv/kg \cdot h$ in birds on high NaCl diet, and to 28 ± 7 $\mu equiv/kg \cdot h$ in birds on medium diet. The Na transport was even in birds on medium NaCl diet restored by aldosterone to only two-thirds of that of birds on low Na diet (Fig. 6.6). The Cl transport was restored to only half that value.

Thomas and Skadhauge (1979b) used the same total dose of aldosterone injected 60 $\mu g/kg$ for 2 days before the perfusion experiment, and observed here also an increase in absorption of Na and Cl, and augmented secretion of K, as compared with controls on low Na diet. Cloacal transport rates were, however, smaller than those attained at the "plateau" level after acute injection (see Fig. 6.11). In the birds receiving a medium NaCl diet the net rate of Na transport was only half that of birds on a low NaCl diet, while in birds on a high NaCl diet it was only one-third. As shown by a Hofstee-plot, aldosterone induced in birds on medium NaCl diet saturation kinetics when the Na absorption rate was correlated with the luminal concentration of Na. In birds on high NaCl diet and chronic treatment with aldosterone there was still a linear relationship between the rate of Na transport and the luminal concentration of Na. It is interesting that the apparent K_m of aldosterone-treated birds on medium NaCl diet was not different from that of birds on low NaCl diet. Injection of aldosterone also enhanced the absorption of Cl and the secretion of K so that the net flow of these ions remained proportional to the rate of absorption of Na.

The effects of acute injections of aldosterone on the electrical transport parameters in isolated coprodea and colons of white Plymouth Rock hens on a commercial diet were investigated by Thomas et al. (1980), who mounted the intestinal preparations in the Ussing-chamber in vitro and measured SCC and PD,

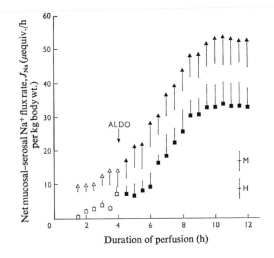

Fig. 6.11. Effect of acute injection of aldosterone on sodium absorption in coprodeum and colon of the domestic fowl. The coprodeum and colon of birds receiving medium *(triangles)* and high *(squares)* NaCl diet were perfused. Aldosterone was injected intravenously after 4 h of perfusion. Values before injection are indicated by *open symbols*, values after injection by *closed symbols*. *Vertical lines* indicate one S.E. The hormone augments the net Na absorption rate. Reproduced from Thomas et al. (1979)

and calculated the resistance. The birds were decapitated 5 h after injection of 128 μg/kg or 32 μg/kg aldosterone. In other experiments 128 μg were administered in four doses over the foregoing 20 h. The coprodeum epithelium was not only exposed to the standard Krebs-phosphate buffer containing 140 mM NaCl but also to a medium containing 6.3 mequiv/l Na (choline chloride replacement). This concentration was chosen since in other experiments (Lyngdorf-Henriksen et al. 1978) it resulted in half maximal flow rate of Na from lumen to plasma in coprodeum of birds on a low NaCl diet. Since the back flux of Na is small, the SCC was also half maximal at this concentration of Na. The purpose of this part of the investigation was to ascertain whether aldosterone influences maximal transport-capacity (V_{max}) or the affinity (K_m) of the entry step at the luminal membrane. A much higher K_m of 51 mequiv/l was observed for inflow across the luminal membrane in birds on a high NaCl diet (see p. 119). In addition to measuring the electrical parameters as such, the electrical reaction of the colon to addition of amino acids, and of the colon and coprodeum to inhibition by amiloride on the mucosal side, was tested. In the coprodea and colons of birds on low NaCl diet the SCC was suppressed by amiloride, and the SCC was not affected by addition of amino acids. In the colons of birds on high NaCl diet the SCC was increased by amino acids, but now unaffected by amiloride. The results of these investigations by Thomas et al. (1980) indicate that aldosterone augments SCC and PD, but not to the values measured in low NaCl diet birds (Fig. 6.12). On the colon aldosterone effected a partial reduction of amino-acid stimulation and a partial induction of inhibition by amiloride. It appears that 128 μg/kg aldosterone results in larger changes than 32 μg/kg. In coprodeum the effects are similar in the medium of low NaCl concentration. The quantitative effect of a high dose of aldosterone is similar on the electrical parameters to that of the reaction to amino acids and amiloride, a 50% effect compared with the low NaCl diet.

The conclusion from these in-vivo and in-vitro experiments is that aldosterone exerts an action on the epithelium of the coprodeum and colon in a manner at least

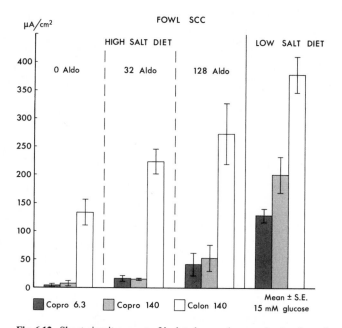

Fig. 6.12. Short-circuit current of isolated coprodeum and colon from the domestic fowl on high and low NaCl diet. Effects of two dose levels of aldosterone (32 and 128 μg/kg body weight) are shown. Coprodeum was incubated in both a low Na medium. (6.3μequiv/l) and a high Na medium (a Krebs-phosphate buffer, Na 140 μequiv/l); colon only in the high Na medium. Aldosterone only restored the SCC partially toward that of preparations from birds on low NaCl diet. Reproduced from Thomas et al. (1980)

qualitatively similar to that caused by NaCl depletion. Why it is quantitatively inferior remains to be elucidated. Three possibilities can be considered: (1) Aldosterone may act synergistically with other hormones that may be released under NaCl depletion. Angiotensin for example acts on mammalian intestine. (2) Aldosterone may require a circadian rhythm in a certain phase with that of other hormones. (3) Non-endocrinological factors such as decreased extracellular fluid volume (ECV) may eo ipso change intestinal ion transport. The ECV affects small intestinal transport in mammals.

Chapter 7

Function of the Salt Gland

The function of avian salt glands is now well known, although development of the present knowledge spans only two decades. Before the description by Schmidt-Nielsen et al. (1958) of excretion of a concentrated NaCl solution from the beak of a double-crested cormorant following a salt load, there were only anatomical descriptions of the supraorbital or nasal glands. The correlation of marine habitat and development of the supraorbital glands was, however, noted by Technau (1936), and the growth of these glands after salt loading of ducks was observed by Schildmacher (1932). The excretion of salt when totally marine birds were feeding on invertebrates was a puzzle because this diet contains a hyperosmotic salt load. Krogh (1939; p. 169) discussed this problem and concluded that these birds must have an efficiently concentrating kidney.

The discovery of the function of the salt glands not only solved the problem raised by Krogh, but also explained the dripping of fluid from the beaks of marine species observed in the past by scientists and naturalists (see Peaker and Linzell 1975; p. 5). The extra-renal secretion through the salt glands, which resembles the function of gills of marine teleosts, presents a challenge of a quantitative description of the function of this organ and interaction with other organs in osmoregulation. Relevant questions include not only its rate of secretion, the nature of the adequate stimulus and the magnitude of change necessary to induce secretion, but also the fraction of a salt load that is excreted by the glands in relation to that removed through the kidneys and cloaca. Finally, as suggested by Schmidt-Nielsen et al. (1963), can a wider interaction among kidney, cloaca, and salt gland be envisioned in the ultimate conservation of water? This would involve continuing excretion of nitrogen waste through the cloaca and NaCl through the salt gland after the cloacal inflow from the kidney of uric acid, salt, and water, and subsequent resorption of salt and water in the cloaca.

Since reviews on function of the salt gland have been published recently (see p. 2), this chapter will consider primarily the *quantitative* role of the salt gland in avian osmoregulation. Shorter sections on other aspects of the topic will serve as introduction. For a comprehensive survey of the literature the reader should consult the monograph of Peaker and Linzell (1975). The interactions of kidney–cloaca and salt gland will be treated in Chap. 8.

7.1 Development of the Gland

The nasal glands, which differ from the lacrymal and Harderian glands, are paired glandular structures with ducts opening into the nasal cavity. Functional salt glands have been described in approximately 50 species of about 20 orders predominantly

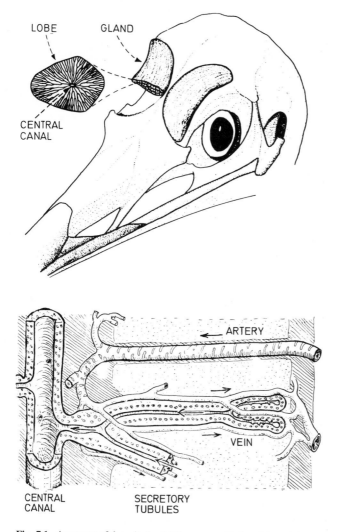

Fig. 7.1. Anatomy of the salt gland. The supraorbital position of the salt gland of the herring gull (*Larus argentatus*) *(top)*. The gland consists of longitudinal lobes with a central canal. Secretory tubules with blind ends radiate from the central canal *(bottom)*. The capillary blood flow is counter to that of the secreted fluid. Reproduced with permission from Schmidt-Nielsen (1960) and from Fänge et al. (1958a)

of marine birds. The name salt gland was introduced by K. Schmidt-Nielsen to emphasize its function in excretion of a solution of high NaCl concentration.

The morphology of a typical salt gland, that of the herring gull, is presented in Fig. 7.1. The acini terminate in ducts with secretory cells opening towards the lumen. These cells are extremely rich in mitochondria.

Comparison of species reveals a gradient in the development of the salt gland from coastal to marine species, with the largest glands observed in birds feeding exclusively on marine invertebrates. The gland weights constitute about 0.1%–2%

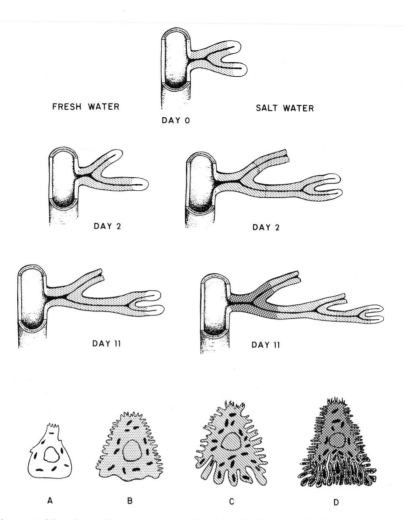

FRESH WATER   DAY 0   SALT WATER

DAY 2   DAY 2

DAY 11   DAY 11

A   B   C   D

Fig. 7.2. Development of the salt gland in response to chronic salt loading. Domestic ducklings received either fresh water or salt-water which resulted in elongation and branching of the secretory tubules. The gland contains peripheral "unspecialised" cells *(A)*, and partially "specialised" cells *(B)* located at the outlet of the secretory tubules. Drinking of salt-water led to full specialisation of the latter cells, first *(C)*, and later *(D)*. Note the basal infoldings and abundant mitochondria localised in the baso-lateral membrane. Reproduced with permission from Ernst and Ellis (1969)

of total body weight (Hughes 1970c). Habituation of individuals to a higher salt intake augments the development of the gland (Staaland 1967; Holmes et al. 1961a), and increases the sensivity and the secretory capacity. This development takes place both by enlargement of the secretory cells, augmentation of the number of mitochondria, and development of the excretory duct system (Fig. 7.2; Ellis et al. 1963; Ernst and Ellis 1969). Concomitantly biochemical changes occur, particularly increases in the activity of the Na- und K-activated ATPase (Holmes 1972).

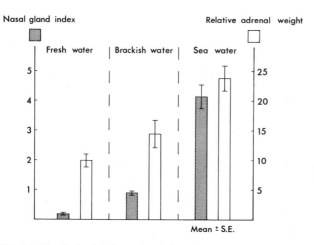

Fig. 7.3. Nasal gland index and relative adrenal weight as functions of salt intake. The relative development of the salt gland (nasal gland index, see text) and the relative weight of the adrenal are both increased in response to the salinity of the environment (eight species in each group). Based on a table from Holmes et al. (1961)

Holmes et al. (1961a) summarised data from eight marine, eight brackish water, and eight terrestrial species as originally observed by Technau (1936), Hartman (1946), and themselves. These data on relative adrenal weights and "nasal gland index" defined as weight of the gland divided by weight of the eye lens, show a remarkable correlation with salinity of the environment (Fig. 7.3).

It should be remembered, however, that functional salt glands are also found in some terrestrial birds of prey (Cade and Greenwald 1966) and in arid-zone birds (Cooch 1964; Ohmart 1972).

7.2 Stimulus for Secretion

The glands receive a rich supply of parasympathetic fibres from the VII (facial) nerve; both electrical stimulation of these fibres and injection of acethylcholine into the blood stream can induce secretion. As with other similarly innervated glands, such as the salivary glands of mammals, secretion is accompanied by a vasodilatation that augments blood flow 14-fold from resting level (Hanwell et al. 1971a). The fractional volume flow is also similar, since about 10% of perfused plasma is secreted in the salt gland of the domestic goose at maximal rate as in the submaxillaris gland of the cat.

The secretion in most birds is either totally inhibited or strongly reduced by anaesthesia or disturbance.

An intact adreno-hypophyseal system is necessary for normal function of salt glands. Salt loading in ducks, induced by maintenance on sea-water instead of fresh water, leads to increased production of corticosterone (see Chap. 9) and injections of both this hormone and of cortisol increased the secretion rate (Holmes 1972, 1978). Conversely, treatment with spironolactone, which inhibits the cellular reaction to several steroids, suppresses the salt-gland response.

There is now a large body of evidence concerning the adequate stimulus for activation of salt-gland excretion, both concerning the peripheral receptors, their sensitivity and specificity, the afferent pathway to the central nervous system, central control, and the efferent pathway. It is now, based mainly on the work on the domestic goose by M. Peaker and associates, established "...beyond doubt..." (Schmidt-Nielsen 1978; p. 302) that the normal stimulus for secretion is an increase of the plasma osmolality, normally the Na concentration, which is registered in receptors in the central vessels, and conducted centrally in the vagus (Hanwell et al. 1972; Peaker and Linzell 1975). Hanwell et al. pointed out that the receptors should be denoted as tonicity receptors since they will react to an increase in plasma osmolality induced by other osmotic agents, such as sucrose, as originally described in the cormorant (Schmidt-Nielsen et al. 1958), and in several subsequent experiments (Peaker and Linzell 1975; p. 46). Occasionally, however, secretion has been observed even after isosmotic volume expansion (Gilmore et al. 1977) (see p. 130).

As most NaCl-loading experiments included an expansion of blood and extracellular volume (see Harris and Koike 1977; Macchi et al. 1967) because the hyperosmotic salt drags water from the cells, it was a question as to whether an increase of the extracellular fluid volume is necessary for activation of the salt gland. That this is not the case was demonstrated with certainty by Stewart (1972) who observed a pronounced secretion in dehydrated domestic ducks that were not salt loaded (see p. 141). These birds had a restricted extracellular fluid volume and decreased plasma volume, but an increase of 11%–13% in plasma osmolality and NaCl concentration. Several observations of encrustations and salts in the nares of dehydrated heat-stressed terrestrial and marine birds have been published (Ohmart 1972; Douglas 1970; Cade and Greenwald 1966; Cooch 1964). These findings also testify to the fact that the salt gland can function without extracellular volume expansion.

The quantitative change necessary to induce secretion varies from species to species and with the length and degree of the salt exposure of the individual bird. McFarland (1964) noticed in 17 species of gulls (including *L. glaucescens*, *occidentalis*, *argentatus* and *californicus*) that small species needed a higher salt load than larger ones. Ash (1969) found that secretion began in the duck (*Aylesbury* drakes) when plasma osmolality and Na concentration was increased by 2%–8%. Peaker et al. (1973) observed that the domestic duck requires an average intravenous infusion of 14.7 mequiv Na/kg body weight to induce secretion, whereas the domestic goose is more sensitive as it requires only 2.0 mequiv Na/kg body weight. The increase in plasma Na concentration at onset of secretion was 6.9%–8.6% in the duck, but only 0%–2.7% in the goose. The most sensitive birds thus require a stimulus—an error signal—no less sensitive than that which induces release of antidiuretic hormone, or elicits drinking response, in mammals. Peaker et al. (1973) also compared the osmotic threshold of domestic geese and ducks to that of gulls (*L. occidentalis*, *glaucescens*, and *californicus*), as investigated by various authors, and the Guam rail (*Rallus owstoni*) and the coot (*Fulica americana*),

(Carpenter and Stafford 1970). Both of these species weigh less than 1 kg. An inverse relationship was found between body weight and mean percentage increase in plasma Na required to initiate salt-gland secretion. Clearly, the sensitivity of the goose and duck makes the salt gland central in ongoing osmoregulation in these species when intake of salt is high, whereas the almost 40% increase required in the coot relegates excretion via the salt gland to an auxiliary or emergency function. More precise information on birds of different orders, from various habitats, and of differing body weights are needed before any conclusions can be made concerning the significance of the suggested correlation. There seems to be no compelling reason to expect body weight per se to be of importance in this respect. Most experiments on secretion by salt glands will not elucidate the problem since the salt load has been beyond the threshold level. The general value of the salt gland in osmoregulation is expressed rather by the ability to tolerate a chronic salt load.

In the individual species the gland develops in response to the daily salt ingestion. This has been investigated especially in domestic ducks maintained on either sea-water or fresh water. W.N. Holmes and coworkers (Holmes 1972; Fletcher et al. 1967) found the following changes on ducks kept on sea-water for 30 days as compared to ducks kept for 30 days on fresh water: The nasal gland weight was augmented by 81%, the ATPase activity by 323%, and the rate of Na excretion by 644%. These relations imply that the ratio between the rate of Na secretion and the amount of ATP hydrolised remained nearly constant. In absolute terms the Na excretory capacity was 22 mequiv/kg day in freshwater-adapted birds and 166 mequiv/kg day in those adapted to sea-water. This excretory capacity thus far exceeds the rate of Na intake which even for a bird drinking 20% of body weight per day of sea-water (see p. 26) is only 90 mequiv Na/kg day. It should be noted further that when the birds are switched back to fresh water, the gland weight, the ATPase activity, and the Na-secretory capacity regress to the previous level. After the first sea-water exposure a further exposure to sea-water switches the responses on much more rapidly. The adrenocortical hormones, which in marine birds are released during salt stress, play a permissive role for this response.

The role of sea-water adaptation was further elucidated by Ruch and Hughes (1975) who loaded freshwater- and sea-water-adapted mallards with 10 mequiv Na/kg body weight intravenously and calculated the salt gland excretion and cloacal discharge for the period 2–3 h thereafter. Their findings in the ducks, and for comparison in the glaucous-winged gull and the male domestic fowl, are presented in Fig. 7.4. It will be seen that the response of the gland is much more rapid after adaptation to sea-water in the mallard, but even in this condition both cloacal and salt-gland excretion are more sluggish than in the gull.

Concerning a possible stimulation via volume receptors and the increase in gland response after chronic salt loading, detailed experiments have been performed on the domestic duck and on the domestic goose.

Zucker et al. (1977) observed in the domestic goose a significant secretion after isosmotic volume expansion alone. They used, however, a larger stimulus (30% of blood volume) than that of Hanwell et al. (1972), who found no effect with an increase of 9%–16% in blood volume. The volume response measured by Zucker et al. (1977) was smaller than that induced by hyperosmotic NaCl loading, since the maximal secretion rate was only 38% of that induced by hyperosmotic NaCl

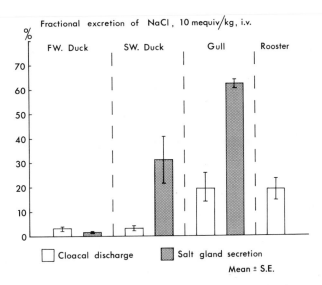

Fig. 7.4. Salt gland excretion and cloacal discharge after a salt load. The salt secretion has been measured for the period 2–3 h after an intravenous NaCl load. The relative rate of the salt-gland secretion of the duck is augmented by sea-water (*SW*) adaptation as compared to freshwater (*FW*) exposure. The gull excretes a higher fraction than the duck. The cloacal contribution is always limited. Based on a table by Ruch and Hughes (1975)

loading, and the average maximal osmolality only 80%, and the Na concentration 68% of that measured after hyperosmotic NaCl loading (see Table 7.1). This difference is consonant with the moderate excretion observed by Stewart (1972) in dehydrated ducks (see p. 141). Peaker (1978) has repeated the experiments of Hanwell et al., but still found no stimulation of the gland after volume expansion.

A volume component is suggested by recent findings of Deutsch et al. (1979), which may also shed light on the different results obtained in the goose. These investigators worked with two groups of domestic ducks, one maintained on fresh water, the other on 2% NaCl as the only drinking solution. This treatment resulted in an average 22 mOs higher plasma osmolality. The birds were subjected to a constant intravenous infusion of a 140 mOs NaCl solution at increasing rates. The results of two typical experiments are reproduced in Fig. 7.5. Three findings are noteworthy: First, in subjects infused with NaCl secretion by the salt gland began much earlier, after an average of 41 min, as compared to 147 min in the non-adapted birds, and after a smaller increase in plasma osmolality, 3.6 mOs as compared to 7.8 mOs. Second, the rate of excretion was much higher in NaCl-loaded birds. The excretion from both salt gland and cloaca was equal to 98% of the injected amount, whereas the non-adapted ducks excreted only 58% of the load by these routes during the infusion period. Third, as the osmolality of the salt-gland fluid (see Table 7.1) exceeded that of the infused solution, the secretion in NaCl-loaded birds was maximal at a time when plasma osmolality was declining. This is in contrast to the increasing plasma osmolality during the entire infusion period in non-adapted birds. This pattern of the NaCl-loaded ducks strongly

Table 7.1. Osmolality, electrolyte concentrations, and secretion rate of salt and fluid. The secretion follows a hyperosmotic salt load, unless otherwise stated

Species	Notes	Osmolality mOs	Na mequiv/l	Cl mequiv/l	K mequiv/l	Flow rate µl/kg min	References
Leach's petrel (*Oceanodroma leucorhoa*)		—	900–1,100	—	—	—	Schmidt-Nielsen (1960)
Black-footed albatross (*Diomedea nigripides*)		—	600–840	590–942	17–22	—	McFarland (1959)
Savannah hawk (*Heterospizias meridionalis*)		—	1,010	1,040	16	100	Cade and Greenwald (1966)
Roadrunner (*Geococcyx californicus*)		1,567	776	794	69	—	Ohmart (1972)
Common puffin (*Fratercula arctica*)		—	—	975	—	196	Hughes (1970b)
Jackass penguin (*Spheniscus demersus*)		1,543–1,580	760–800	666–850	20–29	50–57	Erasmus (1978a, b)
Black-backed gull (*Larus marinus*)		—	784	—	—	227	Schmidt-Nielsen (1960)
Glaucous-winged gull (*Larus glaucescens*)		—	785	872	68	—	Hughes (1970a)
Herring gull (*Larus argentatus*)		—	775	—	—	200	Douglas (1970)
Herring gull (*Larus argentatus*)		—	777	—	26	152	Macci et al. (1967)
Gulls (*L. argentatus and fuscus*)	Mixed	962	727	—	36	28	Ensor and Phillips (1972a)
Brown pelican (*Pelecanus occidentalis*)		—	698	722	13	106	Schmidt-Nielsen and Fänge (1958)
Black swan (*Cygnus atratus*)		—	656	—	19	130	Hughes (1976a)
Cormorant (*Phalacrocorax auritus*)		—	529	517	12	73	Schmidt-Nielsen et al. (1958)
Domestic duck		—	613	610	10	—	Fletcher and Holmes (1968)

Table 7.1 (continued)

Species	Notes	Osmolality mOs	Na mequiv/l	Cl mequiv/l	K mequiv/l	Flow rate µl/kg min	References
Domestic duck		—	—	625	—	68	Scothorne (1959)
Domestic duck	Freshwater-adapted	765	588	—	—	—	Deutsch et al. (1979) and Ruch and Hughes (1975)
Domestic duck	Saline-adapted	1,001	845	—	—	—	Stewart (1972)
Domestic duck	Dehydrated	—	578	—	17	—	Lanthier and Sandor (1973)
Domestic duck		—	531	520	16	151	Carpenter and Stafford (1970)
American coot (*Fulica americana*)		—	530	542	—	142	Gill and Burford (1967)
Domestic goose		—	664	—	15	22	Hanwell et al. (1971a)
Domestic goose		—	430	442	12	71	Zucker et al. (1977)
Domestic goose		864	506	—	—	35	
Domestic goose	Volume-expanded	774	353	—	—	88	Gilmore et al. (1977)
Domestic goose		924	444	—	—	193	Hughes (1968)
Common tern (*Sterna hirundo*)		—	508	—	—	—	
Pintail (*Anas acuta*)		—	550	—	18	—	Cooch (1964)
Gadwell (*Anas strepera*)		—	590	—	20	—	
Brolga (*Grus rubicunda*)		—	263	—	8	47	Hughes and Blackman (1973)

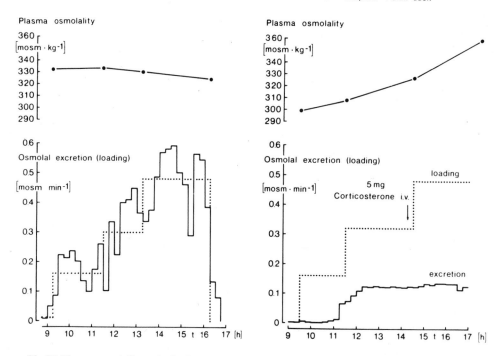

Fig. 7.5. Plasma osmolality and salt-gland excretion in domestic ducks. Salt-adapted birds received 2% NaCl as drinking fluid, the non-adapted tap water. During experiments the birds were loaded by continuous intravenous infusion of a hyperosmotic saline solution. The salt-adapted birds will appear to be able to excrete this load quantitatively whereas the non-adapted cannot; the plasma osmolality of the non-adapted birds rises consequently. Reproduced with permission from Deutsch et al. (1979)

suggests a non-osmotic component in the stimulus to secretion. By venesection and reinfusion of the blood, Deutsch et al. further substantiated the notion that a volume factor was involved. The secretion rate of the salt gland dropped considerably in the NaCl-loaded birds after withdrawal of blood. It returned to previous rates after the reinfusion. It is tempting to speculate that the difference between the volume sensitivity in the findings of Zucker et al. (1977) and those of Hanwell et al. (1972) may be caused by a difference in NaCl in the diets; a contributing factor might be temperature-induced differences in fluid intake.

The induction of secretion by dehydration and heat stress in terrestrial birds (see p. 129) may not be only an osmotic response, since it is possible that hypothalamic temperature per se may play a role. Simon-Oppermann et al. (1979) observed in the domestic duck that hypothalamic cooling leads to a pronounced reduction of the rate of secretion. The cooling induced at the same time a large diuresis. Although hypothalamic warming was not reported during measurements of salt-gland secretion, it is possible to suggest that hypothalamic stimuli as such may contribute to activation of the gland. This may explain the secretion in terrestrial birds of prey, which may be without a hyperosmotic stimulus (Cade and Greenwald 1966). It is

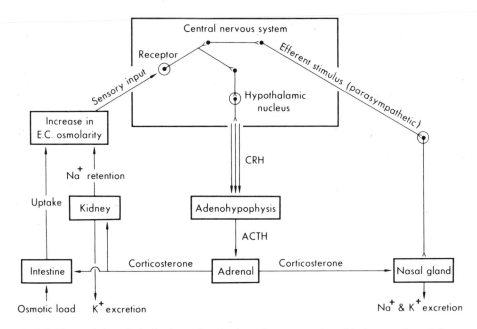

Fig. 7.6. The regulation of salt gland secretion. A schematic representation of the integrated regulation of salt gland secretion. Reproduced with permission from Holmes (1975)

noteworthy that hypothalamic warming was observed to induce a slight antidiuresis in the experiments of Simon-Oppermann et al. (1979).

In summary, the integrated stimulation of salt-gland function involves first and foremost plasma osmolality, second, plasma volume, third, the habitual intake of food and fluid (particularly of NaCl) of the bird, fourth, hypothalamic or other neural factors. Finally, the hypophysio-adrenocortical axis must be intact. The initiation of salt-gland secretion and interaction with other organs has been summarised diagrammatically by Holmes (1975; Fig. 7.6).

7.3 The Flow Rate and Ionic Composition of the Salt-Gland Fluid

The composition of the salt-gland fluid and its flow rate varies with a number of factors, the most important being the species involved, and the salt adaptation of the individual bird. In addition, the ionic composition varies with the rate of secretion. The majority of the approximately 50 species from which fluid has been collected have been examined in only one investigation, whereas the domestic goose and domestic duck, and some species of gulls, have been used repeatedly by several groups of investigators. A much larger body of evidence is consequently available from these species. In most of the other species only concentrations of Na and Cl or K, but not flow rate, have been reported for a few individuals.

135

Table 7.2. Solute composition of the salt-gland fluid from the herring gull. From Schmidt-Nielsen 1960). Unit: mequiv/l

Na:	718	Cl:	720
K:	24	HCO_3:	13
Ca + Mg :	2	SO_4:	1

Common to all birds with a functional salt gland is the excretion of a watery fluid within minutes after an acute load of NaCl. The fluid is two- to sixfold hyperosmotic to plasma. It contains always nearly equal concentrations of Na and Cl, which constitute the majority of the osmotic space. The remainder is made up of K and other ions present in very low concentrations (Table 7.1). In the herring gull (see Table 7.2) Schmidt-Nielsen (1960) found NaCl to constitute 97% of the total measured ionic concentration. Table 7.1 presents concentrations of Na, Cl or K from a number of species. The concentrations of Na and Cl are nearly always at least 20 times greater than that of K. It seems no exception to the rule that the concentrations of Na and Cl are equal, within 10%, and constituting the major fraction of the total osmolality. The only exception seems to be a statement (Schmidt-Nielsen et al. 1963) concerning some samples from the African ostrich in which "...potassium was five to ten times as high as sodium...". For other reasons this species should be reinvestigated (see p. 93). The concentration of K, as of NaCl, may be related to flow rate. Hughes (1970b) observed in the glaucous-winged gull a decrease from 100 mequiv/l to 35 mequiv/l when the flow rate was changed from 0.02 to 0.40 ml/min.

Secretion rates range from 20 to 200 μl/kg min, and the maximal Na and Cl concentrations of acutely NaCl-loaded birds were in the majority of species within 600 to 850 mequiv/l. The K concentration is rarely outside the range of 10 to 25 mequiv/l. The total osmolality, measured in only a few investigations, ranges from 1100 to 1600 mOs. The most often used osmotic challenge in these investigations was a 10 to 30 mMol, NaCl either given orally or injected intravenously in about a 10% solution. In some experiments sea-water or artificial sea-water was used for oral loading. The species that produce a salt-gland fluid of lower or higher ionic concentrations than the range mentioned above do not show a uniform ecological or taconomic distribution, but it is noteworthy that the albatrosses, *Diomedea immutabilis* and *nigripes*, studied by McFarland (1959) all concentrated well, and the terrestrial birds studied by Cooch (1964) concentrated poorly. In his survey Schmidt-Nielsen (1960) noted a correlation for marine birds between the usual range of Na concentrations of the gland fluid and feeding habits of the birds. At the lower end of his scale the double-crested cormorant secretes a fluid of 500 to 600 mequiv Na/l. This species eats fish with a total salinity close to plasma osmolality. The best concentrating species, Leach's petrel *Oceanodroma leucorrhoa*, concentrates to approximately 1000 mequiv/l. It spends most of its life at sea feeding on crustaceans with a body-fluid composition equal to that of sea-water. A further clue to the understanding of a possible correlation between structure and function as related to ecology, in other words gland development in relation

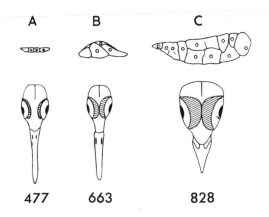

Fig. 7.7. Development and concentrating ability of the salt gland in relation to habitat. The size and location of the glands, and cross section through the glands are shown for **A** the terrestrial wood sandpiper, *Tringa glareola*; **B** the coastal common sandpiper, *Actitis hypoleucos; and* **C** the marine little auk, *Plautus alle*. Both gland size and concentrating ability (Cl concentration, mequiv/l) are correlated with habitat. Modified and reproduced with permission from Staaland (1967)

to concentrating ability, was found when more closely related species were examined. In an anatomical study on 21 species of Charadiiformes (waders, gulls, and auks) Staaland (1967) found that the nasal gland ranged from 0.1% to 1.2% of body weight with a pronounced relation to habitat. The terrestrial birds were at the lower end and the marine birds feeding on invertebrates at the upper end of the weight scale. Staaland also observed a good correlation between the number of lobes and weight of the gland and radius of the lobes (which is approximately equal to the length of the secretory tubules) and marine adaption; and finally a good correlation between length of the secretory tubules and the Cl concentration of the fluid (in 14 species). The data for three selected species are shown in Fig. 7.7. The influence of flow rate on concentration was also demonstrated by Hughes (1972a) who compared spontaneous secretion of the glaucous-winged gull, average 2.1 ml/day, to secretion after a hyperosmotic salt load, 7.8 ml/day. The osmolality was 879 mOs in the former, 1292 mOs in the latter case. It should further be noted that not only the NaCl concentration of the secretion, but the flow rate, also in comparison to the gland weight, increases with size of the glands. This holds well both when species are compared and when saline adaption takes place in individual birds. The concentrating ability seems clearly related to length of the secretory tubules. As the anatomy would not seem to allow any counter-current multiplier principle to function (see Fig. 7.1), the formation of concentrated fluid is most likely caused by a linear concentrating mechanism.

7.4 Mechanism of Secretion

Organ secretions among vertebrates are usually formed by two-stage processes, first, an isotonic fluid is made which, secondly, is rendered hyper- or hypo-osmotic by modification in a duct system. A priori the physiologist would thus seek in the salt gland for a mechanism for formation of an isotonic fluid with another site rendering this fluid strongly hyperosmotic to plasma. As in the salivary glands, there is in the salt glands no anatomical basis for an ultrafiltration process like that of the kidney. There is an anatomical difference between the relatively unspecialised cells at the end of the secretory tubules and the cells lining the ducts which have basal infoldings and many mitochondria so characteristic of ion transporting cells (see Fig. 7.1 and 7.2). The ouabain binding is also maximal by these cells (Hossler et al. 1978). An isotonic primary section may be formed at the closed end of the secretory tubule and modified as it flows along the tubule. Such a model readily explains a correlation between concentrating ability and length of the tubule. There are, however, two possibilities for the formation of a concentrated fluid in the duct system. One is that NaCl is secreted into the tubule across an ion-impermeable apical membrane (Peaker and Linzell 1975; p. 125; Ernst and Mills 1977), the other is that water is withdrawn by a solute-linked water flow across the tubular cells (Ellis et al. 1977). Solutes are in the latter model assumed to recycle from the serosal surface of the cell to the lateral intercellular spaces dragging water across the tight junction and being again extruded on the serosal side. None of these suggested mechanisms are as yet either proven or disproven. The finding of ouabain-inhibited Na-K-activated ATPase, localised to the basolateral membrane (Ernst and Mills 1977) and not to the apical membrane, does not necessarily rule out the model of ion secretion. This location has been suggested in the rectal gland of elasmobranchs (Silva et al. 1977). The high NaCl concentration of the final fluid makes, however, the operation of such a mechanism difficult with passive ion movement across the apical membrane. The water-reabsorption hypothesis is also troubled by quantitative problems. Since the maximum rate of secretion is around 10% of the blood flow to the gland, it does not seem possible to augment this flow fivefold; this would be required to reach the necessary water reabsorption. It should be noted finally that the evidence as to whether the gland cells are hyperosmotic to plasma or not is conflicting (Hokin 1967; Peaker 1971; Schmidt-Nielsen 1976).

The influence of rate of secretion on the ionic concentrations have been studied in different species. The main conclusion is that higher flow rates lead to slightly higher NaCl concentrations. Early investigations pointed, however, to a relative independence between rate of secretion and ion concentrations. Schmidt-Nielsen (1960) observed in the great black-backed gull, *Larus marinus*, an independence of Na concentration and rate of secretion over a range of 400%. Later investigations, particularly those that included secretion at very low rate have, however, revealed a correlation. Hanwell et al. (1971) observed in geese an increase in Na and Cl concentrations from 400–450 to 475–525 mequiv/l when the rate of flow increased from 33 to 100 μl/kg min, but the conclusion did not hold for individual birds. Smith (1972), on the contrary, obtained highly significant positive correlations in individual Aylesbury ducks in which ion concentrations increased from 450 to 600 mequiv/l for a fourfold increase in secretion rate. He noted that in some birds the Cl concentration was higher at higher plasma Cl concentration, but the effect was usually less than 5%. Since, however, most workers have made their observations after single injections of NaCl, the plasma hypertonicity and NaCl concentrations are clearly changing during the response. Finally Deutsch et al. (1979) in their detailed investigation in the domestic duck (Fig. 7.8) found a typical saturation curve relating osmolality of gland fluid to the flow rate. It is clear from their results that

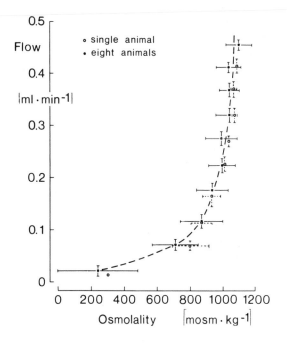

Fig. 7.8. The relationship between flow rate and osmolality of gland fluid. The data are from salt-adapted domestic ducks stimulated by intravenous infusion of hyperosmotic saline solution. The osmolality shows typical saturation kinetics as function of flow rate; but the osmolality is only for very low flow rates significantly reduced. Reproduced, with slight modification, with permission from Deutsch et al. (1979)

investigations using supramaximal osmotic stimuli will work at fairly high rates of flow and thus be near the saturation level at which concentration is indeed flow-independent. The low concentration at low rate of flow, as seen in the domestic duck, is probably parallel to the finding of particularly low NaCl concentrations in some species stimulated only by dehydration or heat stress (see Table 7.3). Some of the rare observations of high K concentrations were also made under such conditions (see Table 7.1). Furthermore, as mentioned previously (see p. 136), Hughes (1970c) observed the highest K concentration at the lowest flow rate. It is not yet possible to interpret the mechanism behind the secretion rate/concentration relationship.

7.5 Adaptation and Hormones

The response of the salt gland of birds exposed to a chronic salt load, as studied in geese, ducks, and gulls, involves both anatomical and physiological adaptations: the weight of the salt glands is nearly doubled (Schmidt-Nielsen and Kim 1964; Ellis et al 1963; Fletcher et al. 1967), the secretory tubules are elongated and become more branched and the secretory cells develop extensive folding of the basolateral membrane, and the number of mitochondria increased (see Fig. 7.2). In a study by Fletcher et al. (1967) the salt load caused a 27% increase in Na and Cl concentration of the fluid, from 466 and 461 mequiv/l to 592 and 585 mequiv/l, respectively. The maximal excretory rate of fluid was tripled. These two changes therefore quadrupled the maximal excretion rate for Na and Cl. The salt adaptation is consequently of great survival value for the bird. Both neural and hormonal responses are necessary for the adaptive hypertrophy after chronic salt loading. Hanwell and Peaker (1975) made unilateral denervations of the gland nerves in geese and observed on the intact side that the RNA concentration was increased in birds that received a hyperosmotic saline as the only drinking fluid for 24 h. In those that received fresh water denervation produced no response. Due to the short duration of the salt exposure no effect on gland weight was to be expected and was not found. The effects of longer exposure should be investigated. In the birds of Hanwell and Peaker (1975) corticosterone injections produced no hypertrophy. The work of Holmes (1972, 1978; p. 130) show, in agreement with this observation, that the adrenocortial and hypophyseal hormones exert a permissive action that modulates the nervous reflex leading to the secretory response after osmotic challenge. Recent investigations have confirmed selective binding of corticosterone to the cytosol of salt-gland cells of the domestic duck (Allen et al. 1975b; Sandor et al. 1977).

7.6 Quantitative Role of Salt Gland in Excretion of an Acute Salt Load and Relation to Kidney/Cloaca in Salt/Water Balance

As indicated by Fig. 7.4, the fractions of injected NaCl that is excreted by the salt gland and by the cloaca may vary considerably from species to species, and with the degree of salt adaptation. This is also clear from several studies on the domestic duck by Holmes and co-workers (Holmes 1972; Holmes 1978).

The capacity of the salt gland usually allows the elimination of 50%–90% of the NaCl load within a few hours. The total excretion is distributed with about three-quarters through the salt gland, the remainder through the cloaca. The species with the higher NaCl concentration (see Table 7.2) of the gland-fluid would be expected to excrete a higher fraction of NaCl through the gland. Comprehensive accounts of salt and water excretion in a few species exist. Data from acutely salt-loaded domestic duck (Smith 1972) and great black-backed gull (Schmidt-Nielsen 1960), dehydrated domestic duck (Stewart 1972), and accipiter hawk feeding on mice (Cade and Greenwald 1966) are presented in Table 7.3. The table clearly shows several aspects of the excretion pattern: First, the importance of the salt gland for elimination of most Na(Cl); even in the non-salt-loaded, but only dehydrated state 75% of Na is excreted through the gland. Second, the relatively low fractional loss of water through the gland unless—during a large salt loading—the secretory rate is very high. A huge salt load also leads to a high renal–cloacal loss of Na resulting in a total water loss that is nearly equal to the amount of water given with the salt when it is presented as a sea-water load. The data from the dehydrated duck and the accipiter hawk may be more representative of the day-to-day role of salt-gland excretion, whereas the NaCl loading experiments probably are characteristic only for sea birds feeding on invertebrates. Third, estimates of rate of evaporation presented by both Cade and Greenwald (1966) and Stewart (1972) show that the

Table 7.3. Elimination of sodium and water through salt gland and cloaca

Species	Stimulus	Salt gland					Cloaca					References
		Water		Sodium			Water		Sodium			
		ml	%	mequiv	%	mequiv/l	ml	%	mequiv	%	mequiv/l	
Black-backed gull (*Larus marinus*)	Salt loading	56.3	43	43.7	91	776	75.2	57	4.4	9	59	Schmidt-Nielsen (1960)
Domestic duck	Salt loading	59	48	35.3	85	598	63	52	6.2	15	98	Smith (1972)
Domestic duck	Dehydration for 1½ days	11.5	11	6.6	75	578	90	89	2.2	25	24	Stewart (1972)
„Accipitrine hawk"[a]	Feeding on mice	1.1	11	1.1	62	1,000	9	89	0.7	38	75	Cade and Greenwald (1966)

[a] Typical values for different species

141

accipiter hawk excreted 60% of water through evaporation, 5% through the salt gland, and 35% from the kidney, while the duck 72% as evaporation, 3% from the salt gland, and 25% through the kidney. It thus appears that under these "normal" functional states of the glands at least two-thirds of Na is excreted by the glands but accompanied only by a very small amount of water. It thus seems relatively unimportant whether the NaCl concentration in the gland fluid is slightly higher or lower. This is in contrast to the case if a large load of sea water is regularly ingested. Here the concentration is important. If, for example, the fluid reaches 960 mequiv/l as compared to the sea-water concentration of 480 mequiv/l, only half of the water ingested with the sea-water is lost through the salt gland, whereas if a concentration of only 270 mequiv/l is reached, three-quarters of the ingested sea-water is lost as gland fluid and only one-quarter available as "free water". It should be noted that during a NaCl load the excretion of K through the gland is not insignificant. Wright et al. (1966) followed excretion by the salt gland and cloaca for 2 h after a single intravenous NaCl load in the domestic duck, which resulted in a total rate of excretion ten times higher for Na than for K. The gland was responsible for the excretion of 36% of K and 78% of Na.

The role of the salt gland can finally be demonstrated by the consequence of extirpation and exposure of the birds to hyperosmotic drinking solution. Ducks thus treated survive only 3–4 days and succumb in a state of dehydration, weight loss, and plasma hyperosmolality (Bradley and Holmes 1971). As the renal concentrating ability in the duck (see p. 68) is less than half that of the salt gland, the glandless bird passes very rapidly into negative water balance if the salt load should be excreted renally.

Extirpation of the supraorbital glands and exposure of the birds to moderate salt loads will, however, also reveal the osmoregulatory value of accessory "salt-gland tissue" and of the tears, the fluid excreted through lacrimal and Harderian glands. Hughes and co-workers have dealt with this problem in several experiments (Table 7.4). The Na concentration was slightly hyperosmotic to plasma with a variable K concentration. In the ducks the tears were responsible for the excretion of an important fraction of both Na and K: around one-fifth of Na and one-half of

Table 7.4. Cation concentration of tears

Species	Na mequiv/l	K mequiv/l	References
Glaucus-winged gull (*Larus glaucescens*)	170–190	1–2	Hughes (1969)
Domestic duck [a]	189	34	Hughes and Ruch (1969)
Domestic duck [b]	142	58	
Brolga (*Grus rubicunda*)	141–180	14–28	Hughes and Blackman (1973)
Black swan (*Cygnus atratus*)	146–198	17–24	Hughes (1976a)

[a] Sea-water
[b] Freshwater

K (Table 7.5). The values must be regarded as approximate since the data were compiled from three sets of observations. In both the kittiwake, *Rissa tridactyla*, (Hughes 1972b) and the glaucous-winged gull (Hughes 1977), birds with supraorbital glands removed tolerated nearly the same chronic salt load as intact birds. This appears to have been caused largely by regeneration of accessory "salt-gland tissue" located over the palate. The duck lacks this regenerative ability (Bradley and Holmes 1971).

Table 7.5. Relative role of tears, as compared to salt gland and cloaca in cation excretion in the salt-adapted duck. (Data compiled by Hughes and Ruch 1969)

	Na %	K %
Tears	21	52
Salt gland	57	9
Cloaca	22	39

Chapter 8

Interaction Among the Excretory Organs

In this chapter the quantitative interaction of kidney, cloaca, and salt gland will be discussed. For the kidney and cloaca the problem is whether the storage of ureteral urine in the cloaca will modify the excretion rates of salts and water. The problem will be treated separately for three major osmotic situations: hydration, salt loading, and dehydration. Urine flow in the water- and salt-loaded states is large compared with the transmural transport in the cloaca, and approximate calculations will be sufficient. In the case of the dehydrated state, however, the ureteral flow of water and salt is so small that the transmural transport in the coprodeum and large intestine will change the rate of excretion considerably. Therefore, more precise calculations must be made. The problem for birds with salt glands will be treated separately. The major question is whether the salt gland and cloaca interact in conservation of water from ureteral urine.

8.1 Birds Without Salt Glands

As the parameters of cloacal transport have been investigated in vivo by perfusion experiments (see Chap. 6; p. 107) in the domestic fowl, the Australian galah, and the emu, and because in-vitro experiments have been carried out on the former two species (see Chap. 6; p. 115), it is possible to quantitate the function of the cloaca in postrenal modification. The assessment of postrenal modification during water and salt loading has been based on the domestic fowl, the only species for which data are available; dehydration can be considered in a general way with computer simulations (Skadhauge and Kristensen 1972). Various strategies for reno-cloacal interaction can be outlined depending on the renal concentrating ability.

8.1.1 Water and Salt Loading

During water loading in the domestic fowl the mean flow rate of ureteral urine is about of 18 ml/kg · h of an osmolality of 115 mOs (Skadhauge and Schmidt-Nielsen 1967a). In the hydrated state the mean osmolalities of the contents in the lower and upper halves of the coprodeum-colon are about 143 and 192 mOs, respectively (Skadhauge 1968). As the mean plasma osmolality is 296 mOs in the hydrated state, and an osmotic permeability coefficient of 3.2 μl/kg · h mOs was calculated (Bindslev and Skadhauge 1971a), the water absorption from the contents in the cloaca may be calculated to be 0.41 ml/kg · h. This is only 2.3% of the flow rate of urine. Although the data on which this approximate calculation is based come from three groups of experiments, it demonstrates beyond doubt that only a minor fraction of the water of ureteral urine is "lost" back into the organism in the hydrated state.

In the salt-loaded domestic fowl the mean flow rate of urine is about 11 ml/kg h with a Na concentration of 150 mequiv/l (Skadhauge and Schmidt-Nielsen 1967a) and with 116 and 94 mequiv/l in the lower and upper halves, respectively, of the coprodeum-colon (Skadhauge 1968). The Na absorption resulting from these concentrations of the mucosal side ranges from 70 to 20 μequiv/kgh depending on the salt loading of the perfused birds (Skadhauge 1967; Bindslev and Skadhauge 1971b; Thomas and Skadhauge 1979a). Since approximately 20% of the NaCl load is excreted in the first hour, about 3 mequivalents, the resorbed amount only represents 0.7 to 2.2% of the amount excreted through the ureters. The possible cloacal resorption is thus quantitatively insignificant for birds on a high NaCl intake.

There is no reason to expect that other species have a different fractional resorption of water and salt in the cloaca when given high water or salt loads. Birds such as salt-marsh sparrows with a chronically high intake of hyperosmotic saline may, however, encounter osmotic problems due to water loss in the cloaca since water cannot be absorbed from a hyperosmotic saline solution when only little NaCl is absorbed. Such species should be studied further.

8.1.2 Dehydration

What happens to the hyperosmotic urine as it flows retrogradely into coprodeum and colon? Both in the large seed-eating species (domestic fowl and turkey) and the smaller (the galah and several other Australian species) and very small forms (e.g. zebra finch) uric acid is deposited on the outside of a central faeces cor (see p. 69) and urine is moved backwards by reverse peristalsis (see p. 34). Although most of the precipitate of uric acid and urates is deposited in the coprodeum the liquid urine moves even into the caeca (see p. 97). The urine is originally somewhat hyperosmotic depending on renal concentrating ability (see Table 5.5), and the ionic composition involves not only NaCl, but also a high concentration of K, NH$_4$, and PO$_4$ (see Table 5.6). Will the storage of this fluid lead to osmotic flow of water across the wall into the gut lumen? Is this water lost by defecation? Or is it possible to envision that urine is first diluted in the coprodeum by absorption of ions and osmotic inflow and then, further retrogradely in the colon both ions and water absorbed? This process, which is illustrated in Fig. 8.1, could result in a net absorption of water. In the colon water would be absorbed from the solution close to the mucosa. Although this solution is still hyperosmotic, solute-linked water flow may overcome the osmotic difference. In the coprodeum and colon of the domestic fowl Skadhauge (1967) showed solute-linked water flow to proceed against an osmotic difference of 65 mOs. The fluid close to the mucosa need not have the same osmolality as the solute fraction of the renal faecal core. Evidence indicates the occurrence of unstirred layers with local low concentrations (Bindslev and Skadhauge 1971b; see p. 109).

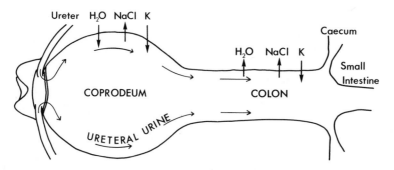

Fig. 8.1. Retrograde flow of ureteral urine into the lower intestine in birds. The ureteral orificia open into the urodeum, from which urine runs into the coprodeum and colon. The urates and uric acid being present in super-saturated colloid suspension are in seed-eating birds deposited outside a central faeces core. The hyperosmotic ureteral urine, formed when the bird is dehydrated, is diluted in the coprodeum. In the colon the osmolality is so low that a net water absorption can take place. Reproduced from Skadhauge (1977)

The problem of retrograde flow has been treated by computer simulation by Skadhauge and Kristensen (1972), using data from the domestic fowl and also from the budgerigar (Krag and Skadhauge 1972). In the model (see Fig. 8.1) the retrograde flow of ureteral urine along the mucosa of coprodeum and large intestine was simulated, with the composition of the fluid at any point along the length of the gut being determined by the combined forces of transmural osmotic water flow, transmural ionic (NaCl) movement, and solute-linked water flow. As the total osmolality does not consist entirely of NaCl, the presence of inabsorbable solutes was assumed. The absorption of NaCl as a function of its luminal concentration was governed by saturation kinetics. Several modifications of the model were explored and all simulations gave essentially the same results as the main model. This shows that the model is applicable to species in which the cloacal morphology and physiological conditions and transport parameters are rather different from those of the domestic fowl.

The principal inference from the model is that the hyperosmotic urine formed in the dehydrated state can indeed be stored in the cloaca without a net water loss from plasma. A small absorption of water (14% of the ureteral excretion) occurs, but with a concomitant absorption of a high fraction of excreted Na(Cl) (Table 8.1). Most values presented in Table 8.1 are based on the dehydrated domestic fowl. If, however, the cloacal transport parameters of the hydrated fowl are used (see p. 112), a 6% loss ("secretion") of water from plasma would occur. The adaptation of the cloaca to dehydration thus saves 20% of the ureteral water output.

That the model probably approximates the true events in the living bird is suggested by two lines of evidence: First, the fractional water absorption was measured in in-vivo perfusion experiments in which the flow rate was as low as that of ureteral urine in the dehydrated fowl. These experiments (Bindslev and Skadhauge 1971b; p. 109) yielded the same results as the model. Furthermore, experiments with exteriorisation of the ureters (Dicker and Haslam 1966) resulted in nearly the same water loss as calculated by the model (Skadhauge 1973; p. 53).

Table 8.1. Computer simulation of cloacal absorption of ureteral urine. (From Skadhauge and Kristensen 1972; and Krag and Skadhauge 1972)

Domestic fowl	Ureteral urine		Fractional absorption	
	Flow rate ml/kg h	Osmolality mOs	Water %	Sodium %
Dehydration	1.0	520	+13.9	+78
Hydration	1.0	520	− 6.1	+55
„Normal case"	1.0	520	+ 5.1	+69
100% increase in flow rate	2.0	520	− 9.5	+39
10% fall in osmolality	1.0	470	+20.3	+78
Non-drinking budgerigar (*Melopsittacus undulatus*)	1.2	720	+ 3	+74
Higher osmolality	1.2	760	− 8	+73

Second, the sensitivity of the model to the various parameters of the system was tested (see below), and the model was most sensitive to the precisely determined parameters (most important the urine osmolality), and least sensitive to the less precisely determined parameters, such as solute-linked flow of water. It should finally be noted that in the experiments of Bindslev and Skadhauge (1971b) raffinose was used as unabsorbable osmotic substance. The fact that ammonium phosphate and not raffinose is naturally present in high concentration in avian urine makes little difference, since phosphate is not and ammonium only slightly absorbed in the domestic fowl (Skadhauge and Thomas 1979). Furthermore, ammonium phosphate has little effect in vitro on Na absorption in the galah (Skadhauge and Dawson 1980b).

Table 8.1 shows the consequence of augmenting the rate of urine flow by 100% and the effect of reducing the urine osmolality by 10%. Both of these changed the fractional water absorption by approximately 15%. The water absorption is thus much more dependent on urine osmolality than on flow rate of ureteral urine. The relative importance of the parameters of the model have been examined systematically and expressed as percentage change in fractional water absorption induced by a change of 10% of each parameter (Table 8.2). The table reveals the paramount importance of urine osmolality. The quantitative difference between the influence of osmolality and flow rate of urine appears to lead to the paradoxial consequence that the bird should conserve water better in the cloaca if the kidney concentrates less well. in other words, if the urine flow increases proportionally with the reduction in osmolality, as it does with unchanged solute excretion rate. The question is whether the augmented cloacal water absorption can overcome the higher renal output of water. What is the combined effect? Calculations of Skadhauge and Kristensen (1972) show that the total water excretion increases with increasing renal concentrating ability, but only slightly, and reaches a plateau at the maximal urine osmolality. If one considers that solute excretion rate (and the urine flow rate) actually decreases with progressive dehydration due to declining GFR, the total conservation of water will increase at higher urine osmolality. Parameters of ureteral and cloacal transport are thus in the domestic fowl precisely matched to

147

Table 8.2. Changes in fractional water absorption in the domestic fowl induced by a 10% increase in parameters shown. (From Skadhauge and Kristensen 1972)

Parameters	% change
Ureteral	
Osmolality	− 13
Flow rate	− 4.2
Cloacal	
V_{max} for Na(Cl) absorption	+ 3.7
K_m for Na(Cl) absorption	− 1.8
Solute-linked water flow	+ 2.2
Osmotic permeability coefficient	− 0.3

+ Denotes increased absorption from lumen to plasma

permit maximal conservation of water and salt, as conditions require. In addition the kidney can regulate excretion of water and salt independently of cloacal storage when the intake of either water or salt is high.

Calculations based on renal parameters of the dehydrated budgerigar and cloacal transport parameters of the domestic fowl corrected for body weight show that this species would loose water from ureteral urine in the cloaca even if the Na concentration were high. (Krag and Skadhauge 1972). Cloacal perfusion experiments with the galah, which also concentrates well, showed, however (Skadhauge 1974b) a similar fractional absorption of water as in the dehydrated domestic fowl: 10%–20% of water, 70% of Na (see Skadhauge 1977). Although several cloacal transport parametrers including K_m and V_{max} for NaCl transport, were "weight for weight" identical with those of the domestic fowl (Skadhauge and Dawson 1980b), the solute-linked water flow was higher, and the osmotic permeability coefficient was lower. This permits the galah to move hyperosmotic urine into the cloaca without further loss of water in spite of the higher urine osmolality.

The emu has an average osmotic urine to plasma ratio of only 1.4 in the dehydrated state (see Table 5.5). The question is whether the cloacal transport parameters are especially adapted to receive this fluid. This question can be answered positively. The absorption capacity of coprodeum and colon is three to six times higher for Na and four to ten times greater for water than in the galah or domestic fowl (Skadhauge et al. 1980). This wide range is due to the effects of varying concentrations of other ions (K, NH_4, PO_4) in the perfusion solution (see p. 114). The water absorption is higher because the solute-linked water flow is closer to an isosmotic solution in this species (see p. 112). Based on cloacal perfusion with a Ringer-like solution, the cloacal absorption capacity is actually larger than the ureteral flow rate of NaCl and water. Too much emphasis can, however, not be placed on the ureteral excretion rates in the experiments of Skadhauge et al. (1980) since the birds were in a state of anaesthetic diuresis. The main conclusion, however, is that also in this species the function of kidney and cloaca seems to be correlated to permit a maximal conservation of salt and water.

The wider aspect of this reno-intestinal exchange of salt and water is that it allows continued nitrogen excretion and avoids clogging of the ureters with colloidal uric acid. Whether the cations trapped in the precipitates of uric acid and urates are recovered in the cloaca remains to be investigated.

Further investigations should be carried out on species with especially large cloacae and caeca (see p. 93), and the parameters of cloacal transport should be measured in species that concentrate well and receive a high salt load, such as salt-marsh sparrows.

On the basis of the evidence at disposal three strategies of reno-cloacal interaction seem to be available to birds: (1) That of the domestic fowl and the galah with fairly high renal-concentrating ability, with cloacal transport matching the incoming fluid. This results in a high absorption of salt and moderate absorption of water in the cloaca. (2) That of the emu with poor renal-concentrating ability, but an almost mammalian-type colon that resorbs large fractions of ureterally excreted salt and water. (3) The yet uninvestigated pattern of good concentrators with a high salt intake. In these species the cloaca may perhaps function as a bladder, being impermeable to minimise water loss. As indicated above salt-marsh sparrows, and also birds of prey and the zebra finch should be studied further from this aspect.

8.1.3 Fractional Absorption/Secretion of K, NH_4 and PO_4

In only one study (Skadhauge and Thomas 1979) has the fractional absorption in the coprodeum and colon of K, NH_4, and PO_4 been quantitatively assessed. In experiments with the domestic fowl these authors observed that the cloaca modifies ureteral urine by secretion of 20% of K and absorption of 8% of NH_4 and 2% of PO_4. The cloaca thus seems to aid in secretion of K, but the storage of ureteral urine affects NH_4 and PO_4 transport only to a small extent. A fairly high rate of absorption of all three ions occurs in the emu (see p. 111), but due to the physiologically less favourable conditions under which quantitative urine collections were made it is premature to calculate fractional absorption rates.

8.2 Birds with Salt Glands

The hypothesis of Schmidt-Nielsen et al. (1963; p. 125) requires that the cloacal absorption of Na and solute-linked water flow remain high when the birds receive a high NaCl intake. As pointed out in Chap. 6 (p. 118) salt loading strongly reduces coprodeal absorption of Na in the domestic fowl. However, since the Na transport by the large intestine remains high in the domestic fowl in salt-loaded state (Lind et al. 1980a; p. 120), and in the domestic duck (B.G. Munck and E. Skadhauge, unpublished observations, the hypothesis of Schmidt-Nielsen et al. may still hold. The large intestine of the domestic fowl does require presence of glucose and amino acids to function in vitro (see Chap. 6; p. 117; Fig. 6.8), but as glucose is needed mainly as a source of metabolic energy (in vitro) (Choshniak et al. 1977),

149

Table 8.3. Excretion of NaCl, and water balance, in ducks with and without salt glands. (From Holmes 1975.) Load: 1.01 sea-water (1,000 mOs), 1 osmol

Excretion	Gland not present	Gland present
Kidney	2.5 l (250 mOs) 1 osmol	0.25 l (250 mOs) 0.1 osmol
Salt gland	0	0.60 l (1,500 mOs) 0.9 osmol
Water balance	loss: 1.5 l	gain: 0.15 l

and because the K_m for amino acid stimulation of Na transport is of the order of a few mM/l (Lind et al. 1980a), the relatively high concentration of glucose and amino acids of chyme from the ileum (see Table 3.1), and the amino acid concentration of ureteral urine of 2 mM/l (p. 72) will always permit stimulation of this Na pump. Further experiments should be carried out in birds with salt glands.

Solutes are thus excreted by kidney, reabsorbed with water by the cloaca and finally secreted by the salt gland. This would conserve "free water" to compensate for the losses, and permit continued excretion of nitrogen. The presence of Na (and K) in ureteral urine may be important to permit precipitates to remain in supersaturated colloid suspension (see p. 84), and some water is needed to wash the precipitate out of the ureteral system. The presence of a salt gland in marine birds that feed on hyperosmotic invertebrates may thus be a better physiological strategy than a higher renal concentrating ability.

As cloacal transport parameters have not been measured in detail in any species with salt glands, a precise assessment of the role of the three organs that excrete salt and water—kidney, cloaca, and salt glands—cannot be made; the relative role of the kidney and cloaca as compared to salt gland has been discussed in Chap. 7.

Finally, the role of the salt gland in comparison with the kidney and cloaca may be illustrated by an example in the domestic duck (Holmes 1975) in which the water requirements for birds with and without salt gland have been exemplified (Table 8.3). It is a clear conclusion that ducks cannot use sea-water without a functional salt gland.

Chapter 9

A Brief Survey of Hormones and Osmoregulation

The most important hormones involved in avian osmoregulation besides AVT, which is treated fully in Sect. 5.2, are those of the adrenal cortex. In this chapter a brief survey on the pituitary-adrenal axis of birds will be presented followed by a short description of the role of its hormones in regulation of renal function. Effects on cloaca and salt gland are treated in Chaps. 6 and 7. A comprehensive survey of the function of adrenocortical hormones has been published by Holmes and Phillips (1976).

9.1 Corticosterone and Aldosterone

The major adrenocortical hormones produced by birds are corticosterone and aldosterone (Sandor et al. 1976). Recent determinations of the plasma concentrations of these hormones are summarised in Table 9.1.

As pointed out by Holmes and Phillips (1976), the published values of the corticosterone concentration in peripheral plasma in birds have been declining over the years because of the steady improvement of specificity of the methods. Fluorimetric methods in particular seem to respond to unspecific material in spite of separation procedures. Recent values recorded for birds not subjected to specific stimulation or stress seem to be in the range of 0.2 to 2.0 μg corticosterone/100 ml plasma. A recent determination of plasma aldosterone was 5.8 ng/100 ml in domestic fowl on a commercial ration (Thomas et al. 1980). These recent values correspond well with earlier double-isotope-derivative assays of adrenal venous effluent from the domestic fowl, which were 6.6 μg corticosterone/100 ml and 0.21 μg aldosterone/100 ml (Taylor et al. 1970). The plasma concentration of corticosterone or corticosterone-like substances is not constant. Both diurenal and seasonal changes of concentration have been observed. In the domestic fowl (Resco et al. 1964) and in the white-crowned sparrow (*Zonotrichia leucophrys pugetensis*) (Wingfield and Farner 1978) winter values were different from spring values. And in the Japanese quail (Boissin and Assenmacher 1968) and pigeon (Bouillé and Boylé 1978) the daily cycle involves two maxima, one in the early morning, the other in the late afternoon with more than 100% difference between the highest and the lowest values.

Adrenalectomy as observed in domestic ducks (Phillips et al. 1961; Thomas and Phillips 1975a) and domestic fowl (Brown et al. 1958b) results in renal loss of NaCl, depressed food intake, and eventual death with haemoconcentration,

Table 9.1. Concentration of corticosterone in avian plasma

Species	Corticosterone µg/100 ml plasma	Method	References
Domestic duck	19.4	Fluorimetry	Machi et al. (1967)
Herring gull (*Larus argentatus*)	5.2	Fluorimetry	
European quail (*Coturnix coturnix*)	2.7	Fluorimetry	Boissin and Assenmacher (1968)
Domestic duck	3.1	Fluorimetry	Bradley and Holmes (1971)
Domestic duck	3.6	Fluorimetry	Holmes et al. (1972)
Domestic duck	1.6	Competitive protein-binding radioassay	Allen et al. (1975b)
Domestic duck	6.4	Fluorimetry	Bhattacharyya et al. (1975a)
Japanese quail (*Coturnix coturnix*)	5.5	Fluorimetry	Bhattacharyya et al. (1975b)
Pigeon	4.5	Fluorimetry	
Domestic fowl	2.0	Fluorimetry	Culbert and Wells (1975)
White-crowned sparrow (*Zonotrichia leucophrys pugetensis*)	1.5	Competitive protein-binding radioassay	Wingfield and Farner (1978)
Domestic turkey	0.2	Radioimmunoassay	Simensen et al. (1978)
Domestic turkey	2.4	Competitive protein-binding radioassay	Krista et al. (1979)
Domestic fowl	0.17	Competitive protein-binding radioassay	Thomas et al. (1980)

hyponatriaemia and hyperkalaemia. This response is identical to that of mammals. A high intake of NaCl or treatment with corticosterone permits continued survival.

Salt loading produces a hypertrophy of the adrenocortical tissue as judged by light- and electron-microscopy (Bhattacharyya et al. 1975a; Allen et al. 1975b).

Salt loading also increases the secretion rate of corticosterone (Donaldson and Holmes 1965), but results in variable changes of the plasma concentration of corticosterone, e.g. either an increase (Taylor et al. 1970), a decrease (Bhattacharyya et al. 1975b), or no change (Thomas et al. 1980). Adrenalectomy affects both the renal and the extrarenal response to a hyperosmotic NaCl load. The onset of secretion by the salt gland is delayed, and the secretory flow almost abolished; the cloacal discharge is also strongly reduced (Fig. 9.1).

The metabolism of corticosterone and of aldosterone, including plasma turnover rate, has been studied extensively in the domestic duck. The clearance of the tritiated compounds was studied by Helton and Holmes (1973), a biological halflife of approximately 30 min was measured for both hormones by Holmes et al. (1974), whereas Thomas and Phillips (1975e) found between 8 and 9 min for both hormones.

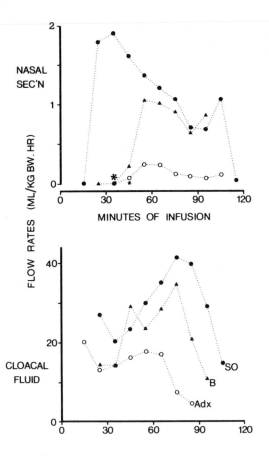

Fig. 9.1. Response of salt gland, kidney and cloaca to adrenalectomy. The temporal courses of responses to an intravenous salt load in sham operated (*SO*), and adrenalectomised ducks with (*B*) and without (*Adx*) corticosterone therapy. The flow both of nasal fluid (*top*) and of cloacal fluid (*bottom*) is delayed and strongly reduced after adrenalectomy without substitution treatment. Reproduced with permission from Thomas and Phillips (1975d)

9.2 Hypothalamo-Hypophyseal Hormones

The influence of these hormones has been elucidated by classical extirpation techniques, by analysis of gland contents of hormones, and by injection of known hormones. The presence of a hypothalamic corticotropic-releasing hypothalamic factor (CRF) has been suggested both in the domestic fowl (Salem et al. 1970a) and the domestic duck (Stainer and Holmes 1969).

The pituitary gland contains and releases the adrenocorticotropic hormone, ACTH, which plays a major role in controlling the output of corticosterone from the adrenal gland (see Nagra et al. 1963; Boissin 1967; and Bradley and Holmes 1971). Injections of ACTH also induce secretion of significantly more nasal fluid in response to a hyperosmotic salt load in the domestic duck (Peaker et al. 1971). In hypophysectomised domestic fowl the renal effects of ACTH and of adrenocortical hormones were found to be variable (Brown et al. 1958a).

The pituitary level of prolactin was augmented in the domestic duck, in herring, and in lesser black-backed gulls (*L. argentatus* and *fuscus*) in response to a chronic salt load (Ensor and Phillips 1970; Phillips and Ensor 1972). Ensor (1975) has

153

further suggested that prolactin and corticosterone act synergestically, and that prolactin plays a general role in modulating the response to dehydration. It seemed to affect intake of both food and fluid and the rate of urine flow. Further experiments should, however, be undertaken to substantiate these observations. Finally, it may be noted in passing that prolactin is responsible for secretion of the "crop milk" with which all columbiform species feed their newly hatched offspring.

9.3 The Pituitary-Adrenal Axis; Renal and Extrarenal Secretions

Hypophysectomy reduces only partially the renal excretion of water, Na, and K as compared to sham-operated domestic duck (Wright et al. 1966), and GFR is halved (Bradley and Holmes 1971), whereas the salt-gland response is almost abolished (Wright et al. 1966). Similar observations were made after total adrenalectomy in the domestic duck. Phillips et al. (1961) found that the renal excretion of water, but not that of Na, significantly increased, but that the salt-gland secretion is totally suppressed. Injection of cortisol causes a significant reduction of renal water output, and a Na retention, but a large secretion of fluid from the salt gland. In later experiments Holmes et al. (1972) observed that the plasma concentration of corticosterone of adrenohypophysectomised domestic ducks was reduced to less than 10% of the concentration in sham-operated birds. Treatment with ACTH restored both the plasma concentration of corticosterone and the ability to excrete a salt load through the nasal gland.

Thomas and Phillips (1975c) studied renal plasma flow (PAH clearance) and GFR (inulin clearance) in sham-operated and totally adrenalectomised domestic ducks. Both parameters declined slightly as functions of time and both were strongly reduced by adrenalectomy. RPF was reduced from about 28 to 8 ml/kg min, GFR from 1.7 to 0.7 ml/kg min (see Fig. 9.1); the reno-cloacal loss of Na was tripled in the adrenalectomised birds. The sham-operated birds resorbed 99.6% of filtered Na, the adrenalectomised birds only 94.4%.

Injections of both aldosterone and corticosterone reduce the renal rate of Na excretion in domestic ducks (Holmes et al. 1961b; Holmes and Adams 1963).

In summary, the hormones of the adrenal cortex are most important for the NaCl balance in birds; however, they exert a stronger influence on the extrarenal than on the renal excretion. They are released during salt loading, and their permissive action is necessary for optimal functioning of the salt gland. Their regulatory role on kidney and cloaca is to conserve Na. The available evidence is not sufficient to assess the role of these hormones in integrated reno-cloacal Na excretion or in the regulation of K balance.

9.4 A Note on the Renin-Angiotensin System of Birds

The kidney of birds possess the juxtaglomerular apparatus (Ogawa and Sokabe 1971) which is present also in bony fishes and all terrestrial vertebrates (Sokabe and Ogawa 1974). As studied in the domestic duck and the pigeon (Chan and Holmes

1971), and in the domestic fowl (Taylor et al. 1970), there is a renal substance in these species that releases a pressor factor after incubation with plasma. In later work the renin production of the avian kidney has been confirmed and the amino acid structure of angiotensin I and II identified (Nakajima et al. 1978; Nishimura 1978). Angiotensin II of the domestic fowl contains eight amino acids. It is identical to the bovine angiotensin II, and it deviates only in one amino acid from the human hormone (Nakayama et al. 1973; Sokabe 1974). Both juxtaglomerular apparatus and renin activity of the kidney are augmented after Na depletion as are the concentrations of plasma corticosterone and aldosterone, but the renal pressor system is not in itself steroidogenic (Taylor et al. 1970).

Infusion into the renal portal system of the domestic fowl of an angiotensin II deviating in one amino acid from the native hormone cause diuresis and natriuresis in spite of a decreased GFR. The effects are more pronounced on the side of the infusion (Langford and Fallis 1966). Angiotensin II may thus have both glomerular and tubular actions in the control of the function of the avian kidney.

It has been demonstrated beyond doubt that the renin-angiotensin system is involved in regulation of blood pressure of birds and in the drinking response (see p. 14). To what extent it participates in intrarenal regulation (tubulo-glomerular balance), activates other hormones (aldosterone), or affects epithelial transport remains to be elucidated.

Chapter 10

Problems of Life in the Desert, of Migration, and of Egg-Laying

10.1 Desert Birds

In a hot and dry environment birds must master the two major physiological problems of heat dissipation and water conservation. These problems are related, as water is used when animals are forced to resort to evaporative cooling. Successful adaptation to the desert therefore not only involves water conservation by the kidney and other organs, but also the control of evaporative cooling (see p. 51). This involves setting of critical temperature, level of heat loss by radiation, convection and conduction, and adjustment of motor activity. Some species of mammals increase body temperature during day-time, whereby heat is stored and evaporative cooling is reduced or avoided. The excess heat is then released largely as radiation during the night. By a hypothalamic mechanism panting is suppressed in such animals when they are dehydrated. Respiratory water loss may also be a regulated parameter through an adjustment of H_2O lost per unit uptake of O_2. This effect may be caused either by more efficient respiration, or by a counter-current exchange in the airways that lowers the temperature of the expired air (see p. 47).

The studies of desert adaptation have followed the lines of both field observation and laboratory investigation. The first informs us about drinking, feeding and flight patterns, mobility, and reproduction in relation to rainfall. The second involves not only the classical parameters of osmoregulation, evaporation and excretory water loss, but also tolerance to increase in plasma osmolality, metabolic water production, and "fuel" and water economy during flight. The latter parameters determine mobility.

Field observations have shown that desert birds do not exist in the strictest sense, in that no species live exclusively in deserts, or have achieved complete independence of drinking water. Desert adaptation must be viewed as a relative concept in which the species of birds that are widespread in arid zones are those in which important adjustments of the various processes of osmoregulatory importance may be found.

Some exceptions to the dogma that desert adaptation primarily involves water conservation are apparent. Three will be mentioned here. The first is that of birds of prey. In order to exist at all they must feed, and their prey contains 60%–70% water. This is apparently enough to cover their water requirement, as they are rarely observed to drink at desert water sources (Fisher et al. 1972). Their flight altitude will also put them above the heat stress of the desert (air temperature falls ca. 6.5°C/1000 m, and birds have been observed to rise to altitudes around 10 km). The second and third exception concern priority. The need for energy conservation and for thermoregulation may overrule water conservation. The first is seen in black animals in the desert. Their absorption of heat is a disadvantage in summer, but may be of decisive importance in conservation of energy when food is scarce, in black colour in the roadrunner. When the ambient temperature is below its zone of thermoneutrality this bird assumes a "sunning behaviour" that allows maximal exposure of black skin. The average energy saving was calculated to be the equivalence of 41% of the basal metabolic rate for a bird of the weight of the roadrunner. It should be borne in mind, however, that other factors than feather colour play a role in total heat balance. At moderate to high wind velocity the heat load is determined by the balance between penetration of radiation into the erected plumage and convective cooling. Calculations of Walsberg et al. (1978) suggest that dark colour under such conditions actually results in less heat gain than white plumage!

The second exception which concerns priority is seen when dehydration overrules thermoregulation as discussed in Chap. 4 (see p. 51).

Two small species that are both common to the arid interior of Australia, the zebra finch and the grass parrot, the budgerigar and two larks of the Namib desert of South West Africa, Stark's lark and the grey-backed finch-lark have been examined physiologically in some detail (see the following section).

10.1.1 The Zebra Finch

Investigations on zebra finches have been performed both on native West Australian birds and on domesticated North American and European stocks that are an unknown number of generations removed from their native habitat. As to be outlined below, a number of parameters were found to be identical in the two groups, but others differed significantly. It is characteristic for both domesticated and wild zebra finches that some, but not all, individuals can survive for several weeks to months on dry seeds alone under ordinary laboratory conditions of temperature and humidity (Calder 1964; Cade et al. 1965; Oksche et al. 1963; Skadhauge and Bradshaw 1974). Sossinka (1972) had more than 50% survivors after 1.5 years of water deprivation. In the dehydrated state the sum of preformed water in food and metabolic water production just matches the cloacal loss plus the evaporation (Table 10.1).

A second observation to be discussed below is that dehydration or restricted water intake always reduce the measured or calculated rate of evaporation, also in relation to oxygen uptake (Calder 1964; Cade et al. 1965; Lee and Schmidt-Nielsen 1971; Skadhauge and Bradshaw 1974).

Calder (1964) measured water intake, oxygen consumption, and pulmonocutaneous water loss together with cloacal water excretion in native and domesticated finches, but a difference was not observed. Calder found an ad-libitum water intake of 3.0 ± 0.3 ml/24 h equivalent to 24% of body weight. The standard metabolic rate was 3.3 ± 0.1 mlO$_2$/g · h. Both observations were confirmed by Cade et al. who found a water intake of 24% of body weight/day and a standard metabolic rate of 3.7 ml O$_2$/g · h. The thermoneutral zone was measured

Table 10.1. Water budget and other parameters of dehydrated, non-heat-stressed desert birds

	Zebra finch (*Poephila guttata*)	Stark's lark (*Spizocoryx starki*)	Grey-backed finch-lark (*Eremopterix verticalis*)	Budgerigar (*Melopsittacus undulatus*)	Gambel's quail (*Lophortyx gambelii*)
Body weight (g)	13.4[a]	16[g]	16[g]	30[b]	149[h]
H$_2$O intake ad lib. (g/24 h)	3.8	2.3	1.3	1.5	11.2
Food intake g/24 h	3.3	1.9	1.8	4.9	7.0
Preformed water in food (g/24 h)	0.46	0.17	0.16	0.52	0.6
Metabolic water production (g/24 h)	1.57	0.93	0.85	2.30	2.5
Water gain (g/24 h)	*2.03*	*1.10*	*1.01*	*2.82*	*3.1*
Cloacal loss (g/24 h)	(0.61)[c] 0.80[a]	1.00	0.78	0.65[c] 0.91	6.1
Evaporation (g/24 h)	1.20	0.37	0.26	1.86[c] (1.67)[f]	4.8
Water loss (g/24 h)	*2.00*	*1.77*	*1.04*	*2.77[b]*	*10.9*
Water balance (g/24 h)	+ 0.03	− 0.27	− 0.01	+ 0.05	− 7.8
Urine osmolality (mOs)	1,005	—	—	947	669[i] 962[j]
Water content of droppings (%)	65[e]	52	51	60 59[c]	44[h]
g H$_2$O evaporated/l O$_2$ uptake (g/l)	0.54[a] 0.54[d]	0.69	0.47	0.55[b]	0.57

References: [a] Skadhauge and Bradshaw (1974). [b] Krag and Skadhauge (1972) (Ureteral urine). [c] Cade and Dybas (1962). [d] Lee and Schmidt-Nielsen (1971). [e] Calder (1964). [f] Weathers and Schoenbaechler (1976). [g] Willoughby (1968). [h] McNabb (1969b). [i] McNabb (1969a). [j] Carey and Morton (1971).

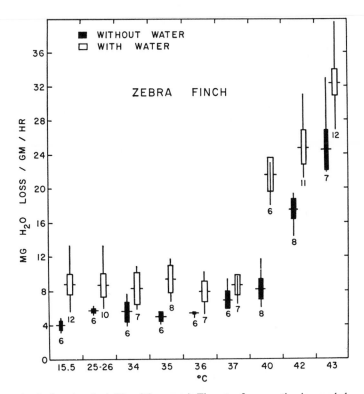

Fig. 10.1. Rate of evaporation in the zebra finch (*Poephila guttata*). The rate of evaporation is recorded as function of the ambient temperature in normally hydrated and in dehydrated zebra finches. Dehydration results in a significant reduction of the rate of evaporation. Reproduced with permission from Cade et al. (1965)

by Calder with 30–40°C, by Cade et al. with 36–40°C. At higher ambient temperature the water consumption increased, but dehydrated or water-restricted birds lost constantly less water than birds with free access to water (Fig. 10.1). Identical observations were made by Calder (1964) and by Lee and Schmidt-Nielsen (1971). The latter authors addressed the question of whether this reduction is due to respiratory or cutaneous conservation of water. They measured either total evaporation or respiratory evaporation by covering the body with a plastic bag. In the latter case identical rates of evaporation were found per unit of oxygen consumption in watered and dehydrated birds indicating that the difference in the dehydrated state was exclusively due to a decrease in cutaneous water loss, the mechanism for which is unknown. When the tolerance to intake of solutions of varying salinity was tested, a marked difference was observed between domesticated and Australian zebra finches. The former had lower concentrations at maximal rate of intake, and lower maximal salinity of fluids the birds would drink (Table 10.2). This difference was also reflected in a higher osmolality and Cl concentration in the fluid in the lower end of the droppings (see Fig. 5.4). This shows that the wild birds have a higher maximal renal concentrating ability. When

159

Table 10.2. Saline intake and renal concentrating ability in wild and in domesticated zebra finches

	Domesticated	Wild
Salinity at maximal drinking rate	0.2 *M* NaCl[a] 0.2 *M* NaCl[c]	0.3 *M* NaCl[d] 0.5 *M* NaCl[b]
Maximal salinity taken in	0.6 *M* NaCl[c]	0.8 *M* NaCl[b, d]
Maximal osmolality of droppings	704 mOs[e]	1,027 mOs[d]
Maximal Cl conc. during intake of saline at maximal rate	224 mequiv/l[c]	318 mequiv/l[d]

References: [a] Cade et al. (1965). [b] Oksche et al. (1963). [c] Lee and Schmidt-Nielsen (1971). [d] Skadhauge and Bradshaw (1974). [e] Skadhauge (1973)

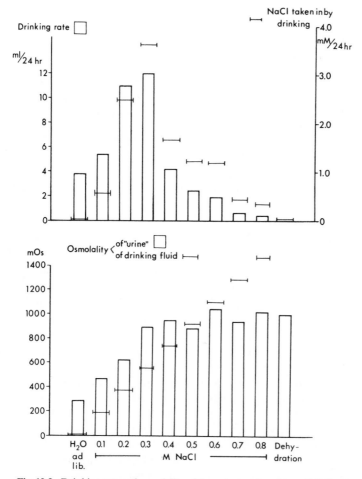

Fig. 10.2. Drinking rate and osmolality of droppings as functions of NaCl content of the drinking fluid. The drinking rate is increased until the osmolality of the fluid taken in approaches that of the droppings ("urine"), but drinking continues even when the drinking fluid (0.8 NaCl) is 40% hyperosmotic to the excreta. Reproduced from Skadhauge and Bradshaw (1974)

160

the intake of saline is examined in greater detail (Fig. 10.2), two aspects become apparent. First, the intake of saline fluids increases until the osmolality of the fluid being drunk approaches the maximal renal concentrating ability. Second, fluids of higher salinity are being drunk. In the first case "free water" is no longer generated since more concentrated fluids cannot be formed by the kidney. The birds would then be expected to stop drinking. It is, however, certain that fluids up to 0.8 M NaCl are actually being drunk, since the NaCl concentration of the droppings are larger when this saline is available than in birds exposed to simple dehydration. The urine osmolality is the same in both situations, ca. 1000 mOs (Skadhauge and Bradshaw 1974). This apparently paradoxical finding prompted a closer examination of the water balance of a non-drinking zebra finch as shown in Table 10.1. For the calculation of the water budget, food intake and its water content was measured directly, and the metabolic water production calculated from the composition of food and standard conversion factors and fractional absorption rates. The rate of evaporation was the minimal value 4 mg/g · h estimated by Cade et al. (1965) (see Fig. 10.1), and the calculation shows that the birds are in water balance. It should be noted that if the birds were assumed to be in water balance, as indeed they are without drinking water when they maintain body weight, the resulting rate of evaporation, 123 mg/24 h, is, as compared to the oxygen consumption calculated from the oxygenation of protein, fat, and carbohydrate, equal to 0.54 mg H_2O/ml O_2. This is identical with the values observed by Lee and Schmidt-Nielsen (1971) (0.54 mg H_2O/ml O_2) and by Cade et al. (1965) (0.56 mg H_2O/ml O_2). Calder (1964) found in his finches an identical food intake, 2.9 g food/ 11.9 g body weight, but a slightly lower cloacal water loss 0.54 g/24 h. Sossinka (1970) maintains, however, that food intake tends to be smaller in domestic zebra finches that show a smaller overall activity.

The important finding of Table 10.1 is that the water turnover is high in the non-drinking zebra finch: 15% of body weight per 24 h. This turnover rate was confirmed by experiments with tritiated water. In these experiments the total body water was determined at 63% of body weight (Skadhauge and Bradshaw 1974). The water turnover is thus 24% of total body water/24 h. The water taken in from solutions of 0.8 M NaCl amounts only to around 10% of the turnover (Oksche et al. 1963; Skadhauge and Bradshaw 1974). It may be speculated that the NaCl aids in water conservation by producing drier faeces. This, however, does not seem to be the case since Lee and Schmidt-Nielsen (1971) observed almost double the amount of water/g dry faeces in birds drinking 0.6 M NaCl than in dehydrated birds. Since the intake of highly concentrated saline solution is so small compared to the total turnover of body water in the dehydrated state, it is perhaps beneficial in maintaining extracellular volume, although augmenting the osmotic stress. Until a precise balance of solute excreted in solution is made, the exact mechanism for osmoregulation during drinking of concentrated solution of NaCl remains unknown.

The ability to drink fairly concentrated saline solutions may have survival value. Although the water of Australian desert water holes even after several months without rainfall rarely contains Na in concentrations higher than 40 mequiv/l (Fisher et al. 1972), Cl concentrations of over 300 mequiv/l have been observed at the peak of the dry season (see Skadhauge and Bradshaw 1974). Finally, breeding is related to rainfall (Serventy and Whittell 1967; p. 8) and to ambient temperature (Davies 1977).

10.1.2 The Budgerigar

The intake of fluid and food of the budgerigar was first measured by Cade and Dybas (1962) as functions of temperature and humidity of the surrounding air, and of the NaCl concentration of the drinking water. The ability of the budgerigar to maintain body weight without drinking water was noted by Cade and Dybas and confirmed by others (Greenwald et al. 1967; Krag and Skadhauge 1972). Concerning intake of saline solution these authors observed that many birds refused the NaCl solutions whereas others, the "drinkers", would consume increasing amounts at higher salinity and often die even when drinking a salinity of only 0.3 M NaCl. The water content of the droppings was also measured (see Table 6.2), and the pulmono-cutaneous water loss was calculated. The birds with the lowest physical activity had an average value of 1.86 g/bird · day. The authors still consider this value high, as they calculated a water gain of only 1.84 g/bird · day from water of food and metabolic water production. As they measured a cloacal loss of 0.65 g/bird · day, only 1.19 g should be available for evaporation. A total water budget of a non-drinking budgerigar was made by Krag and Skadhauge (1972) (see Table 10.1). Their birds had a total water gain of 2.88 g/bird · day which provided both for a rate of ureteral urine flow of 0.81 g/bird · day and an evaporation as large as that determined by Cade and Dybas (1962). The measurement of evaporative water loss was confirmed by Weathers and Schoenbeachler (1976) who found a basal value of 1.67 g/bird · day which is low for a 30 g bird (see Fig. 4.2). There is thus general agreement among the quantitative measurements of various authors of all major parameters of water excretion.

Greenwald et al. (1967) measured both body temperature and evaporative water loss as related to ambient temperature, and they measured the renal Cl excretion when their birds drank up to 0.4 M NaCl solutions. They observed a range of thermoneutrality from 34°C to 44°C. This is high compared with other bird species. Like the zebra finch (see p. 159), the budgerigar reduces evaporative water loss when exposed to dehydration, but only at ambient temperatures above the lower critical temperature.

The plasma Cl concentration was measured during a period of dehydration. The mean plasma concentration of Cl is about 117 mequiv/l in normally hydrated birds; it increases to an average maximal value of 132 mequiv/l after 2 weeks of dehydration, but declines after an additional 3 weeks to the original value. A maximal urine Cl concentration of 445 mequiv/l is reached with drinking of 0.4 M NaCl. Drinking of isotonic saline solutions results in plasma Cl concentrations of about 160 mequiv/l.

The renal concentrating ability of the budgerigar was further explored by Krag and Skadhauge (1972). An average maximal urine osmolality of 848 ± 26 mOs was observed in ureteral urine of dehydrated birds. A slightly higher concentration was observed in the liquid part of the naturally voided droppings: 947 ± 60 mOs. The average osmotic urine to plasma ratio in the former group was 2.3 with a maximal individual value of 3.0. Food intake and rate of urine flow of the water-deprived birds are recorded in Table 10.1.

From these observations the budgerigar emerges as well adjusted to water conservation: (1) Absolute rate of evaporation is low with further reduction during dehydration (mechanism unknown), and tolerance to high ambient temperature before evaporative cooling sets in. (2) High renal concentrating ability. (3) Tolerance to high plasma osmolality and electrolyte concentrations.

In addition to these factors two other aspects related to water conservation have been elucidated. First, energy consumption and evaporation during flight, second, postrenal modification of ureteral urine. Tucker (1968) measured O_2 consumption, CO_2 production, and rate of evaporation in trained budgerigars during flight at several rates of air speed. In addition, temperature dependence and influence of ascending or descending flight were tested. Of osmoregulatory interest was the finding of an oxygen uptake during level flight at 35 km/h of 15.8 1/24 h for a 30 g bird. The respiratory quotient (RQ) was 0.78 indicating oxidation primarily of fat. The rate of evaporation during this experiment at 18–20°C ambient temperature was 14.4 g/24 h · 30 g, whereas the metabolic water production was calculated at 9.4 g/24 h · 30 g bird. Compared to the caged, water-deprived birds (see Table 10.1), the oxygen uptake of these flying birds was augmented 4.6-fold, the metabolic water production 4.1-fold, but the evaporation 7.7-fold. Naturally, these birds could no longer remain in water balance. At this temperature only 15% of heat production was, however, lost by evaporation. At 36–37°C ambient temperature the evaporation increased to 46 g/24 h · 30 g bird, equal to 47% of heat production. The interesting question is, therefore, for how long the budgerigar can fly without access to water before dehydration leads to an intolerable increase in plasma osmolality and loss of extracellular volume? Using Tucker's (1968) and their own data, Krag and Skadhauge (1972) calculated this with 12 h flight or a distance of 420 km. The budgerigar thus seems to have reasonable mobility to find both food and water in the desert. It is, however, also clear that long flights equivalent to migration would require a flight altitude where ambient temperature is lower than 18–20°C. An additional heat load from solar radiation must also be anticipated. That a high flight altitude may be necessary for prolonged migratory flights without access to free water is also the conclusion reached by Torre-Bueno (1978) (see p. 168). In this context it is noteworthy that Tucker (1968) calculated that the energy required for ascent is virtually recovered during descent so a high flight altitude does not present the birds with an extra energy requirement.

Concerning the problem of cloacal resorption of—or water loss from—ureteral urine, computer calculations were made by Krag and Skadhauge (1972). In the calculation the parameters of ureteral urine as determined in the budgerigar were used together with cloacal transport parameters as determined on the domestic fowl on a weight-to-weight basis. The result of the computer calculations was that ureteral urine with a flow rate of 1.2 μl/g · h and an osmotic urine to plasma ratio of 2.3 could flow into the cloaca without a net water loss, but only if a large fraction of Na present in a high concentration in the cloaca were absorbed (see Chap. 8; p. 147). However, since the osmotic permeability coefficient of the cloacal wall may be lower in the budgerigar than in the fowl and the rate of solute-linked water flow higher, as in the galah (Skadhauge 1974b; p. 110), the net result may be that no water loss occurs in the cloaca in spite of the fairly low Na concentration of the cloacal contents.

When finally the budgerigar and the zebra finch are compared, both show desert adaptation by maintaining a high rate of food intake and, therefore, metabolic water production in the dehydrated state, a relatively low rate of evaporation, and formation of a fairly concentrated urine most probably without further cloacal water loss. In addition, the budgerigar has a high critical temperature and exhibits a great mobility by flight without a high water loss. The zebra finch is able to drink fluids of fairly high salinity and its breeding behaviour is linked to rainfall.

10.1.3 Larks from the Namib Desert and Gambel's Quail

Daily water budgets of the non-drinking Stark's lark and grey-backed finch-lark from the Namib desert, and Gambel's quail, have been included in Table 10.1 for comparison with the Australian species. These species have in common a low rate of evaporation and a low rate of water loss by the reno-cloacal route. Furthermore, the rate of evaporation is low in relation to metabolism. The factors that allow the majority of these species to survive on a diet of dry seeds without access to water are thus well established. As in other cases of adaptation the key to success lies not in an outstanding capacity of any particular parameter, but in the total effect of a favourable value within a normal range for other birds of all relevant parameters.

10.1.4 The Ostrich

The water economy of the African ostrich (*Struthio camelus*) has received considerable attention since Fourcroy and Vauquelan in 1811 first noticed the high content of uric acid (17 g/l) in its urine. The urine voided from the cloaca was observed in the dehydrated state to reach a maximal osmolality of nearly 800 mOs (maximal urine/plasma osmotic ratio of 2.7) (Louw et al. 1969). As in the emu (see p. 95), the cloacal anatomy of the ostrich (see p. 93) would seem to allow a high rate of absorption from the ureteral urine. Direct measurements should be carried out. The water content of the droppings was reported to fall by 22.4% from the hydrated to the dehydrated state, but absolute values were not stated (Cloudsley-Thompson and Mohamed 1967). These authors also observed that the ostrich could withstand a weight loss of 25% during dehydration, in the course of which the (cloacal) body temperature increased. This was also observed by Crawford and Schmidt-Nielsen (1967) who found a reduced evaporative cooling as the dehydration progressed (see p. 52). This mechanism aids in water conservation. The african ostrich thus seems well adapted to life in arid lands in spite of its size. In its Australian counterpart, the emu, a low turnover of water (see p. 7) has been demonstrated. This bird differs from the ostrich in having a low renal concentrating ability (see p. 67) and no salt gland. The function of the latter in the ostrich should be investigated.

10.2 Problems of Osmoregulation During Migratory Flights

Little is known about the osmotic problems involved in trans-ocean or trans-desert migratory flight, during which neither food nor water are available. There are numerous investigations of pre-migratory fattening, and measurements of fat content and water content after shorter or longer flights, but only few other measurements pertaining to osmoregulation.

Two questions are of interest: Is there any change in body water or electrolyte content in preparation to flight and during flight, and do the birds arrive dehydrated or with disturbed salt balance? Due to lack of physiological measurements none of these questions can be answered, but on available evidence it is possible to pass judgement whether energy reserves or water reserves limit range of flight. It should be noted that for birds, as also observed for insects (grasshoppers), the storage and metabolism of fat yields a maximum of metabolic water production. This plays a significant role in the maintenance of water balance during migration.

In this section a short survey of observations of water balance after migratory flights is presented first, and then of observations and calculations of the evaporative water loss (particularly from the respiratory tract) due to the increased metabolic activity during flight.

10.2.1 Water Balance During Migratory Flights

This problem has received some attention mostly in small birds weighing around 20 g; larger birds, except the pigeon, have not been investigated. A priori, smaller birds might be considered to face greater problems of osmoregulation since, due to pronounced pre-migratory fattening, they are able to migrate long distances (60–70 h flight) over the open ocean. Larger birds fly predominantly over land at day-time. They often stop en route to feed and drink. They follow coastlines utilising thermic currents which permit soaring flight; this requires less energy than forward flight in still air, which is less economical in larger birds. As forward flight would have required higher heat loss due to evaporative cooling, the conclusion is that if larger birds embarked on prolonged trans-ocean flights they would exhaust both energy resources and water, and they do not. Thus larger birds, in general, avoid osmotic problems during migration.

Small pre-migratory fat-storing birds on the contrary migrate at night, avoiding inflow of heat by solar radiation, and at a slower speed. This requires a rate of metabolism that can be sustained without a high rate of evaporative cooling if the air temperature is favourable. The air temperature is such that the major fraction of heat can be dissipated as radiation and convection. These birds may voluntarily select a lower air temperature by flying at higher altitudes.

In order to elucidate the water balance after migration, birds have been caught before and after migratory flights and the body water content analysed. Due to the large amount of fat, measurements of total body water are misleading as the adipose tissue has low water content. Birds that are caught after a long flight during

Table 10.3. Body water and fat in palm warblers *(Dendroica palmarum)* before and after migration (From Johnston 1968)

Location	Body weight g	$H_2O\%$	Fat %	$H_2O\%$ of non-fat dry weight
Tallahassee (before flight)	11.15	51.7	23.9	67.4
Ship (after flight)	8.40	63.7	5.1	67.2

which they have lost nearly all their fat will therefore have a higher relative water content: One example will suffice: Johnston (1968) compared palm warblers *(Dendroica palmarum)* killed at a Florida television tower (at Tallahassee) to birds that landed on a ship near the North coast of Cuba, thus after a long flight. The essential data are presented in Table 10.3 which shows that the relative water content of the entire body is very low in the premigratory fattened bird. A value of about 50% of H_2O was also found in other species (Moreau and Dolp 1970; Haas and Beck 1979). The total body water increases to "normal" values (cf. Table 1.1) when the birds burn off the fat. If, however, the body water is expressed as a fraction of non-fat dry weight, no difference between before and after flight is apparent (see Table 10.3). It should be noted that the fraction of body water derived from this or similar experiments is not directly comparable to body water measurements in other birds (see Tables 1.1 and 1.2), as the water of the fat is included and feathers, legs, and gut contents often subtracted.

The fat of migratory birds is usually low in water content. McGreal and Farner (1956) studied the lateral thoracic body of the white-crowned sparrow and found that from March to April the relative water content decreased from 30% to about 12% of the weight. Odum et al. (1964) observed on red-eyed vireos *(Vireo olivaceus)* a water fraction in adipose tissue of 40% in the spring, declining to 20% in early fall, and 6% in late fall just before migration. These authors also observed a constant water content in relation to non-fat dry weight in several species of parulid warblers collected at a Florida television tower in spring (after trans-ocean flight) and autumn (before migration). The spring birds had an average water fraction of non-fat weight of 68.2%. This value was unchanged in the autumn (67.1%) although the autumn birds had a 3.2 times higher fat content. The water content has also been measured in European birds migrating across the Mediterranean to Africa and over the Sahara desert. Several species were examined by Moreau and Dolp (1970) on landing in Egypt after trans-Mediterranean flight. The fat content was variable, and the water content related to non-fat weight ranged from 67% to 58%. These birds cannot on the basis of these observations be regarded as dehydrated (cf. Table 1.1). Haas and Beck (1979) who shot and collected exhausted birds in the desert of the Western Sahara (Tanezrouft) and compared them with control specimens collected north and south of the desert, found an average of 65% of H_2O in control birds, 63% in birds shot in the desert, and 55% in exhausted birds caught by hand. The control species were, however, small (7 to 14 g), the shot birds larger (9 to 62 g), but the exhausted birds weighed 64 to 125 g. The difference between the control birds and those shot in the desert was not significant. Thus the conclusion from these investigations is that none of the investigated birds, except the larger hand-caught ones, were dehydrated. Haas and Beck suggest that smaller birds become exhausted because they run out of fat, the larger ones due to dehydration.

It may be concluded that the majority of investigations have failed to reveal convincing evidence of dehydration after migratory flights. The parameter "water per cent of non-fat weight" is, however, rather crude since water of fat tissue is excluded, but gut water, etc. included. Direct measurements of blood osmolality should be carried out. Concerning evidence of dehydration Fogden (1972) felt that the migratory birds become dehydrated when the non-fat water fraction declines to less than 66%. This value is, however, not lower than the water content of other birds (Table 1.1). The above-mentioned experiments express dehydration as a loss from before to after flight. The water loss may also be either calculated or measured directly during flight.

10.2.2 Calculations and Measurements of Energy Expenditure and Evaporative Water Loss During Migratory Flights

Yapp (1956) calculated the metabolic energy required for flight, and concluded that more water was needed for evaporation than was produced by metabolism. Dehydration might therefore limit range of migration. His values both for energy requirement and water loss were, however, high; particularly the latter consideration was based on flight measurements at fairly high ambient temperature. Nisbet et al. (1963) and Nisbet (1963) summarised the literature and made two important conclusions. First, blackpoll warblers (*Dendroica striata*) have enough fat when they leave New England to fly non-stop even to Venezuela. Second, the weight loss during migration is, based on several field studies, 0.6% to 1.3% per hour. The average power consumption for passerine birds was estimated to 88 W/kg body weight. These findings seem to indicate that earlier calculations of metabolism during flight and of evaporative water loss were too high in comparison with the true situation during migration. Early high estimates were based on a larger species, the pigeon, and higher ambient temperature (Zeuthen 1942; Salt 1964). That this bird does require evaporative cooling during flight caused by non-respiratory ventilation was confirmed by Hart and Roy (1966), but the previous estimates of energy expenditure and therefore of evaporation were too high.

Hart and Berger (1972) summarised flight energetics and water economy in migratory birds, including several measurements of oxygen consumption during flight. The oxygen uptake is independent of body weight, and approximately 12 times that of basal metabolic rate. Several indirect estimates on migratory birds indicate the same energy consumption during migration as estimated by Nisbet (1963): 87 W/kg. The problem is whether this rate of metabolism requires more water for evaporative cooling than is produced by the fat consumption. If pure fat is metabolised producing 39 KJ/g yielding 1.07 ml H_2O, and 2.4 KJ is dissipated by evaporation of each gramme of water, only 6.7% of the produced heat can be dissipated by evaporation. Obviously, a small amount of water will be lost from the cloaca, but this is of the order of 1 μl/g $\cdot$ h (see Table 10.1). The metabolic

167

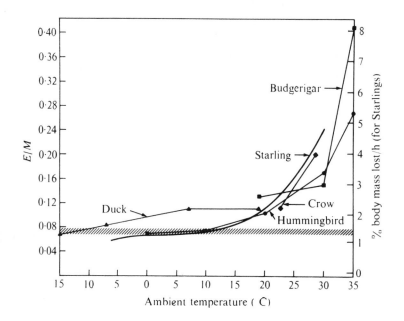

Fig. 10.3. Ratio between evaporative cooling and metabolic rate (E/M) as functions of ambient temperature. The *shaded* line shows the range of metabolic water production depending on the ratio of fat and carbohydrate being metabolised. Evaporation close to or lower than the values expressed by this line thus indicates that the bird can remain in water balance. It will appear that an ambient temperature lower than approximately 12°C is needed for the maintenance of water balance. Reproduced with permission from Torre-Bueno (1978)

production is $\dfrac{0.087 \cdot 3600}{39} \cdot 1.07 = 8.6 \,\mu l/g \cdot h$. The cloacal loss is therefore only 10% to 15% of total water production. The problem is now how low the air temperature must be to permit evaporative cooling dissipate only around 5% of the total loss including radiation and convection. Torre-Bueno (1978) observed in wind-tunnel experiments in two European starlings that evaporative water loss exceeds metabolic water production at an average ambient temperature of 7°C with an upper limit of 12 °C due to experimental uncertainty. This value is strikingly close to the maximal estimate of flight temperature of 13°C made by Nisbet et al. (1963). The findings in the starling and measurements of other birds are presented in Fig. 10.3. Even if the evaporative water loss were higher than metabolic water production, several hours of flight would have to pass before dehydration and a measurable weight loss (Hart and Berger 1972) or increase in plasma osmolality would occur. If 87 W/kg is metabolised equivalent to 8.6 μl H_2O/g h, a 50% higher water loss corresponding to an air temperature during flight of 20°C (see Fig. 10.3) would result in an additional water loss of 4.3 $\mu l/g \cdot$ h. A flight of 20 h would thus result in a water loss of only 86 μl or 8.6% of body weight. The total weight loss would be 16.1% of which 7.5% was fat. This flight would result in an increase in plasma osmolality of approximately 12%. This value has often occurred in dehydrated birds (see Table 1.4) with no ill effect (Skadhauge 1974a).

168

Migratory birds may therefore meet a rather unfavourable air temperature or even sustain solar heat radiation, and the loss of lean body mass will not be detectable due to statistical variation. Plasma osmolality must be determined to reveal this limited dehydration. Note that burning of carbohydrate at 0.6 g H_2O/g will only yield 56% of the water, but require 2.3 times higher fuel weight. As energy is wasted carrying fuel, a hypothetical carbohydrate-loaded bird could only fly one-third of the distance at no more than $-5°C$!

In summary, the question of whether energy reserves or water availability limit range of migratory trans-ocean flights can be so answered: Small migratory birds, which undergo substantial fattening, sustain prolonged flights without serious dehydration. This conclusion is based an two lines of evidence. First, estimate of body water content has not revealed significant water loss after migration. Second, direct and indirect estimates of energy expenditure, metabolic water production, evaporation, and reno-cloacal water loss strongly suggest that migratory birds remain in water balance provided the average air temperature at flight remains about 10°C. Twice this temperature can be tolerated during prolonged flights without serious increase in plasma osmolality. Water is thus not limiting the range of migration. The larger day-migrating birds seem, in general, to depend on soaring flight; they remain in contact with land and avoid dehydration.

10.3 Osmotic Problems of Nestling Birds

Newly hatched birds encounter the same basic problems of water balance as adult birds. Dehydration threatens if the birds receive inadequate intake of water or if excessive evaporation occurs due to thermoregulatory defense. The evaporative capacity has been studied in nestling birds (see Table 4.1) which are generally fully capable of dissipating the heat generated by metabolism through evaporation of water (Bernstein 1971a; Hudson et al. 1974; Dawson et al. 1978). Active function of the salt gland has also been observed in nestling roadrunners in the field (Ohmart 1972).

Species that do not provide food with enough water, regurgitate food from the crop with sufficient water in the juices. In pigeons, doves and other species a nutritive fluid or liquid "milk" is formed from the crop (see Ziswiler and Farner 1972; p. 359). In a few species a peculiar water-transport mechanism has evolved, e.g. the water-transport in sand grouse feathers (see the following section). It may be noted in passing that an osmotic interaction exists between the young and the parents when the latter ingest the faeces and urine voided by the young. Some nestling passerines discharge urine-faeces in a so-called faecal sac, which is often swallowed by the parents. As the young usually discharge a product of lower solute concentration than the reno-cloacal concentrating ability of the adult, the latter may gain some "free" water by feeding on the faecal sacs. For the mountain white-crowned sparrow (*Zonotrichia leucophrys oriantha*) (Morton 1979) and the roadrunner (Calder 1968) the water from the faecal sacs may add a significant amount of water to the daily balance although precise quantitative estimates are not available.

10.3.1 Water Transport by Sand Grouse Feathers

Cade and MacLean (1967) have studied sand grouse in the Namib desert of Southwest Africa and confirmed observations by earlier naturalists that the adult male sand grouse soaks the belly feathers in water for several minutes when he visits a water hole. The water is then carried to the nestling site up to 30 km away. The young gather around the male, taking the abdominal feathers in their beak. Photographic documentation of this behaviour was also produced by George (1969) in the Senegal sand grouse (*Pterocles senegallus*) studied in the Morocco desert. Cade and MacLean measured the total water holding capacity of breast feathers of the four species of Namib desert sand grouse, and compared them with those of several other birds. The other birds held around 5 g H_2O per g dry feather, equivalent to the water holding capacity of ordinary filter paper. The specially adapted belly feathers carry around 12 g H_2O per g in females and 15–20 g per g in males. The highest value was found in the male of *Pterocles namagua*. Thomas and Robin (1977) found even higher values in *P. senegallus* (24.7 g per g) and in *P. lichtensteinii* (26.6 g per g) from the Moroccan desert. These authors also measured the water taken up by the hydrated feathers themselves to be about 5 g per g. The difference between the total water capacity and the hydrated feathers is the usable or "strippable" water. This available water is thus in the best-adapted species around 20 g H_2O per g dry feather. The total water carrying capacity per bird has also been estimated by Cade and MacLean who measured it by soaking the belly feathers of dead specimens in water. They measured the carrying capacity with about 25 ml water in specimens of *P. namagua* weighing 160–180 g. They also tried to calculate the loss by evaporation during flight to the young to be about 50% for a flight of several miles at desert temperature and humidity. The conclusion is that there should after one flight be sufficient strippable water to provide 4–5 nestlings of 20 g with an amount equivalent to 10% of their body weight.

Cade and MacLean also described the special development of the belly feathers. The proximal four-fifths have specialised barbules which when touched by water form a meshwork that holds water by capillarity (surface tension). The meshwork is formed because the barbules, which in the dry state are parallel to the feathers, are uncoiled by hydration of the feather keratin (Rijke 1972; Joubert and MacLean 1973), so that they spread out at right angles to the feathers. It is thus the water taken up by the barbules that permits the holding of a large amount of "strippable" water by capillary force. It should finally be noted that in addition to this special water transport to the young Thomas and Robin (1977) found other signs of xerophilic adaptation in the desert sand grouse. They concentrate virgorous activity to the cooler part of the day, and maintain total immobility and thermal insulation when the ambient temperature exceeds body temperature. They avoid evaporative cooling unless ambient air temperature exeeds 50°C. Finally, the kidney structure suggests a good renal concentrating ability.

10.4 Water and Salt Metabolism in Relation to Egg-Laying

The studies relating osmoregulation to oviposition fall in two groups: The first includes work on hormonal changes during the egg-laying cycle and hormonal induction of oviposition, the second contains observations on drinking, water turnover, and urinary excretion in relation to oviposition.

Ovulation in birds is apparently caused by release of luteinizing hormone (LH), and oviposition appears to be associated with release of AVT. Tanaka and Nakajo (1962) showed that the neurohypophyseal content of AVT underwent a drastic reduction (by 60%) at the time of oviposition in hens. The content of oxytocin was only slightly reduced. The AVT concentration of plasma was assayed by Sturkie and Lin (1966) (see p. 83) just before and during oviposition. Compared to control periods, the concentration was increased 3-fold and 133-fold, respectively. These studies suggest that normal oviposition is associated with release of AVT. Other findings indicate, however, that AVT is not necessary for oviposition, particularly because neurohypophysectomy does not prevent normal egg-laying (Shirley and Nalbandov 1956; Opel 1965). In a later study Sturkie and Lin (1967) showed that oviposition could be induced without release of AVT. Other studies have elucidated the action of AVT on the smooth muscle of the reproductive tract. AVT has a clearcut effect on intrauterine pressure in the hen (Rzasa and Ewy 1971) and on contractility of the oviduct in vitro (Rzasa 1972), and the hormone can also induce oviposition (Rzasa and Ewy 1971; Rzasa 1978). The physiological role is thus far from fully established. The notion that corticosterone induces the release of LH was supported as dexamethasone prevented ovulation in the hen by abolishing the preovulatory raise in LH (Wilson and Lacassagne 1978). LH is, however, only released by corticosterone in pharmacological doses (Sharp and Beuving 1978).

10.4.1 Water Intake

Early investigations showed that laying hens consumed around 1.7 times more water than cockerels (Lifschitz et al. 1967). The water turnover was also determined by the HTO-dilution method by Chapman and Black(1967) who measured the half time (t 0.5) of water in the total body-water pool. The half-time (t 0.5) was defined by the equation t $0.5 = (\ln 2)/k$ in which k is the rate constant, i.e. the slope of the logarithm of the isotope concentration in the pool as a function of time. The water turnover was found to be much higher in the laying hens as t 0.5 was 3.6 ± 0.3 days compared to 7.3 ± 0.3 days in the cocks. The total body water, about 63% of the weight, was not significantly different between the groups. The turnover values are equivalent to a turnover of 125 ± 16 ml H_2O/kg · day in laying hens and 61 ± 1.8 ml H_2O/kg · day in the cocks (see Table 1.2). Essentially the same turnover rates were observed for laying hens and cocks by Chapman and Mihai (1972), who also examined the hens before they started laying. In the non-laying state the turnover of water is only 73 ± 4 ml/kg · day as compared to 122 ± 12 ml/kg · day during laying. It is thus obvious that egg-laying is associated with an increase in water

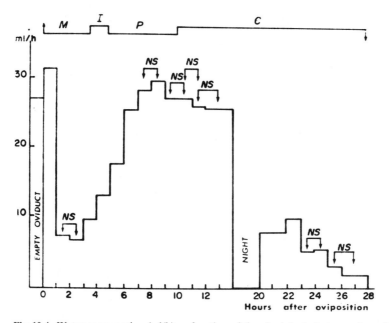

Fig. 10.4. Water consumption (ml/h) as function of the physiological stages of egg formation in the domestic fowl. A three-phasic pattern with wide excursions is observed. The symbols indicate: *M*, Egg in magnum; *I*, Egg in isthmus; *P*, Albumen plumping; *C*, Shell calcification. *NS* indicates a non-significant difference. Reproduced with permission from Mongin and Sauveur (1974)

turnover. The increase is most likely due predominantly to increase in rate of metabolism, whereas the water content of an egg, around 40 ml, is small compared with the increase in the intake of drinking water. This increase was for a 3.5 kg hen 172 ml/day.

Attempts to elucidate further the water intake in relation to oviposition have been made by Mongin and Sauveur (1974) and by Howard (1975), who measured hourly water consumption in relation to the time of oviposition. Both papers measure water consumption as a function of time of the day. A diurnal pattern with no drinking in the dark was very obvious from the study of Howard (1975). Drinking began at "lights on" with a small peak before midday and the largest intake in the late afternoon. This pattern was also found by Mongin and Sauveur (1974) and by Nys et al. (1976). For laying birds both papers normalised water intake to the time of oviposition. Mongin and Sauveur observed a triphasic response: As shown in Fig. 10.4, a very high rate of intake (50 ml/h) was observed for a short interval immediately after oviposition, with the second highest peak (37 ml/h) occurring during albumen plumping and shell calcification, and a third small maximum a few hours before oviposition. The latter two peaks were also noted by Howard (1975). The low rate of drinking measured just before oviposition, and the earlier high level were also noted by previous investigators (Lifschitz et al. 1967) and by Wood-Gush and Horne (1970). Howard (1975) also measured urine flow rate and plasma and urine osmolality and presented results as fractional changes from the values of non-laying hens. The birds were apparently in

172

a state of water diuresis before oviposition as the rate of urine flow was high and plasma and urine osmolality low. After oviposition a marked antidiuresis lasted for approximately 2 h.

The excretion of "strong" electrolytes in relation to egg-laying has been measured by Taylor and Kirkley (1967), who measured 24 h food intake, and faecal and urinary output on egg-laying and non-egg-laying days in four colostemised hens. Egg-laying was associated with a significant retention of Cl and K, but not of Na, whereas a urinary excretion of NH_4 was observed. The intake of food was higher on egg-laying (average 112 g) than on non-egg-laying days (78 g). Mongin and Lacassagne (1967) observed the urinary output of several electrolytes for 5 days before and 3 days after the laying of the first egg. Urine was, however, only collected for 1 h during the dark. Oviposition was associated with retention of Cl, excretion of Na, and little change in output of K.

It must be concluded that there are large increases in food and fluid intake associated with egg-laying, but the NaCl turnover is less affected. It may be noted in passing that there is retention of Ca and HCO_3 and excretion of P. As the interval of egg-laying in the fowl is slightly longer than the daily rhythm, the changes associated with oviposition per se will be difficult to sort out. A strong dependence of salt/water relations on the hormonal changes associated with egg-laying does not emerge. The increased metabolism and water turnover during egg-laying periods is not accompanied by water retention or weight gain (Howard 1975). In conclusion, oviposition seems to be associated with polydipsia of unknown cause and with release of AVT strong enough to cause antidiuresis and contract the oviduct. This response is presumably physiological, i.e. meaning normally occurring, but it does not seem to be necessary for reproduction.

Notes Added in Proof

Since the completion of the manuscript several papers pertinent to osmoregulation in birds have appeared. Papers published up to the end of 1980 are briefly surveyed below.

Chapter 1. Robinzon et al. (1980) reviewed comparative aspects of salt preference and intake in birds; Dantzer and Braun (1980) compared the nephron function in reptiles, birds and mammals, and Sandor and Medhi (1980) surveyed the role of adrenocorticortical stereoids in extrarenal electrolyte secretion in non-mammalian vertebrates.

Chapter 2. Kaufman et al. (1980) found in the domestic pigeon that drinking was induced by paraenteral injection of hyperoncotic solutions of either polyethylene glycol or dextran. To explain this response the authors suggest that drinking receptors could be situated somewhere in the extracellular compartment, perhaps in the brain.

Chapter 3. Morley et al. (1980) observed in the domestic fowl on a high NaCl diet an increased water transport across jejunum and a higher plasma Na concentration. Following injections of ovine prolactin, these authors (Morley et al. 1981) observed a reduced sodium and water transport across jejunum and "rectum". Their interpretation is that prolactin, released by increased plasma Na, reduces intestinal uptake of NaCl, thus acting to maintain homeostasis.

Duke et al. (1981) caecectomised great-horned owls and observed an increased water consumption, which, however, was only pronounced in the second week after the operation.

Chapter 4. Collins et al. (1980) measured evaporative water loss in the singing honey-eater and the brown honey-eater (*Lichmera indistincta*).

173

Chapter 5. Karasawa (1980) following previous experiments on urinary excretion of uric acid in the domestic fowl fed a low protein diet (Karasawa 1977) noted that a high protein diet resulted in a large increase in urinary NH_4 output following an intraveneous ammonia infusion.

A sensitive radioimmunoassay was developed to AVT in serum from the domestic duck (Möhring et al. 1980). The detection limit was 2 fMol. In ducks maintained on fresh water the plasma concentration was 5.1 ± 1.4 pMol/l (mean $\pm$ S.E.), at a plasma osmolality of 294 mOsm. In ducks maintained on saline water (600 mOsm) the plasma concentration was 22.7 ± 3.0 pMol/l. The plasma osmolality in these birds was 333 mOsm. This assay thus allowed an estimate of the sensitivity of the AVT system to change in plasma osmolality. This was around 0.5 pMol AVT/l mOsm. This is close to the values observed in mammals. In a subsequent paper (Simon-Oppermann et al. 1980), this group showed that blockage of the vagus nerves induced, as in mammals, an antidiuresis associated with increase in plasma AVT.

Chapter 6. The cloacal epithelium of the domestic fowl was examined by scanning and transmission electronmicroscopy (Dahm et al. 1980). In agreement with Schrader and Weyrauch (1976) these authors found in coprodeum a villus epithelium with a simple layer of columnar cells with microvilli. These absorptive cells were interspaced by goblet cells.

Chapter 7. Hammel et al. (1980) further examined the stimulus to salt gland secretion in the domestic duck. They found indication of a volume factor in the stimulus as an increase in the extracellular fluid volume decreased the threshold concentration for Na in plasma.

Lingham et al. (1980) confirm that saline treatment augments the synthesis of NaK-stimulated ATPase in the salt gland of the domestic duck; and Wilson and Butler (1980) observed that corticosterone treatment did not change the intracellular Na-concentration.

Chapter 9. In several papers the plasma concentration of corticosterone has been reported in the domestic fowl (Freeman and Flack, 1980; Freeman et al. 1980a; Freeman et al. 1980b; Curtis and Flack 1980) and the domestic duck (Harvey and Phillips 1980; Harvey et al. 1980).

References

Abdel-Malek MT, Huston TM (1975) The effect of environmental temperature and sodium intake on [22]Na retention. Poult Sci 54:624-625

Acher R (1963) The comparative chemistry of neurohypophysial hormones. Symp Zool Soc London 9:83-91

Akester AR (1964) Radiographic studies of the renal portal system in the domestic fowl (*Gallus domesticus*). J Anat 98:365-376

Akester AR (1967) Renal portal shunts in the kidney of the domestic fowl. J Anat 101:569-594

Akester AR (1971) The blood vascular system. In: Bell DJ, Freeman BM (eds) Physiology and biochemistry of the domestic fowl, vol II, Chap 32. Academic Press, London New York, pp 783-839

Akester AR, Mann SP (1968) Adrenergic and cholinergic innervation of the renal portal valve in the domestic fowl. J Anat 104:241-252

Akester AR, Anderson RS, Hill KJ, Osbaldiston GW (1967) A radiographic study of urine flow in the domestic fowl. Br Poult Sci 8:209-212

Allen JC, Abel JH, Takemoto DJ (1975a) Effect of osmotic stress on serum corticoid and plasma glucose levels in the duck (*Anas platyrhynchos*). Gen Comp Endocrinol 26:209-216

Allen JC, Abel JH, Takemoto DJ (1975b) Uptake and binding of labeled corticosterone by the salt gland of the duck (*Anas platyrhynchos*). Gen Comp Endocrinol 26:217-225

Ames E, Steven K, Skadhauge E (1971) Effects of arginine-vasotocin on renal excretion of Na, K, Cl, and urea in the hydrated chicken. Am J Physiol 221:1223-1228

Anderson RS (1967) Acid-base changes in the excreta of the laying hen. Vet Rec 80:314-315

Andersson B (1978) Sodium versus osmotic sensitivity in cerebral control of water balance. In: Jørgensen CB, Skadhauge E (eds) Osmotic and volume regulation. Munksgaard, Copenhagen, pp 84-99

Annison EF, Hill KJ, Kenworthy R (1968) Volatile fatty acids in the digestive tract of the fowl. Br J Nutr 22:207-216

Argenzio RA, Miller N, von Engelhardt W (1975) Effect of volatile fatty acids on water and ion absorption from the goat colon. Am J Physiol 229:997-1002

Ariyoshi S, Morimoto H (1956) Studies on the nitrogen metabolism in the fowl. 1. Separation of urine for the nutritional balance studies. Bull Nat Inst Agric Sci 12:37-43

Aschoff J, Pohl H (1970) Der Ruheumsatz von Vögeln als Funktion der Tageszeit und der Körpergröße. J Ornithol 111:38-47

Ash RW (1969) Plasma osmolality and salt secretion in the duck. Q J Exp Physiol 54:68-79

Assenmacher I (1973) The peripheral endocrine glands. In: Farner DS, King JR (eds) Avian biology, vol III, Chap 3. Academic Press, London New York, pp 183-286

Barlow JS, Slinger SJ, Zimmer RP (1948) The reaction of growing chicks to diets varying in sodium chloride content. Poult Sci 27: 542-552

Barnes EM (1972) The avian intestinal flora with particular reference to the possible ecological significance of the cecal anaerobic bacteria. Am J Clin Nutr 25:1475-1479

Barnes EM, Impey CS (1974) The occurrence and properties of uric acid decomposing anaerobic bacteria in the avian caecum. J Appl Bact 37:393-409

Bartholomew GA (1964) The roles of physiology and behaviour in the maintenance of homeostasis in the desert environment. Symp Soc Exp Biol 18:7-29

Bartholomew GA (1972) The water economy of seed-eating birds that survive without drinking. In: Voous KH (ed) Proc 15th Int Ornithol Cong E J Brill, Leiden

Bartholomew GA, Cade TJ (1956) Water consumption of house finches. Condor 58:406-412

Bartholomew GA, Cade TJ (1958) Effects of sodium chloride on the water consumption of house finches. Physiol Zool 31:304-310

Bartholomew GA, Cade TJ (1963) The water economy of land birds. Auk 80:504-539

Bartholomew GA, Dawson WR (1953) Respiratory water loss in some birds of southwestern united states. Physiol Zool 26:162-166

Bartholomew GA, Dawson WR (1954) Body temperature and water requirements in the mourning dove, *Zenaidura macroura marginella*. Ecology 35:181-187

Bartholomew GA, MacMillen RE (1960) The water requirements of mourning doves and their use of sea water and NaCl solutions. Physiol Zool 33:171-178

Bartholomew GA, MacMillen RE (1961) Water economy of the California quail and its use of sea water. Auk 78:505-514

Bartholomew GA, Hudson JW, Howell TR (1962) Body temperature, oxygen consumption, evaporative water loss, and heart rate in the poor-will. Condor 64:117-125

Baylé JD, Bouillé C (1978) Regulating mechanisms of adrenocorticotropic function in birds. In: Gaillard PJ, Boer HH (eds) Comparative endocrinology. Elsevier, Amsterdam, pp 371-374

Beattie J. Shrimpton DH (1958) Surgical and chemical techniques for in vivo studies of the metabolism of the intestinal microflora of domestic fowls. Q J Exp Physiol 43:399-407

Begin JJ, Johnson TH (1976) Effect of dietary salt on the performance of laying hens. Poult Sci 55:2395-2404

Bell DJ, Bird TP (1966) Urea and volatile base in the caeca and colon of the domestic fowl: the problem of the origin. Comp Biochem Physiol 18:735-744

Bell DJ, Freeman BM (eds) (1971) Physiology and biochemistry of the domestic fowl, vol I, II. Academic Press, London New York

Benoff FH, Buss EG (1976) Water consumption and urine volume in polydipsic and normal white leghorn chickens. Poult Sci 55:1140-1142

Bentley PJ (1971) Endocrines and osmoregulation. A comparative account of the regulation of water and salt in vertebrates. Zoophysiology and ecology, vol I. Springer, Berlin Heidelberg New York, pp 1–300

Berde B, Boissonais RA (1968) Basic pharmacological properties of synthetic analogues and homologues of the neurohypophysial hormones. In: Berde B (ed) Neurohypophysial hormones and similar polypeptides. Handb Exp Pharmakol 23. Springer, Berlin Heidelberg New York, pp 802–870

Berde B, Huguenin R, Sturmer E (1962) The biological activities of arginine-vasotocin obtained by a new synthesis. Experientia 18:1-4

Berger C (1966) Mikroskopische und histochemische Untersuchungen der Niere von *Columba livia abersatio domestica*. Z Mikr Anat Forsch 1:436-456

Berger L, Yu TF, Gutman AB (1960) Effects of drugs that alter uric acid excretion in man on uric acid clearance in the chicken. Am J Physiol 198:575-580

Berman A, Snapir N (1964) The relation of fasting and resting metabolic rates to heat tolerance in the domestic fowl. Br Poult Sci 5-6:207-216

Bernstein MH (1971a) Cutaneous and respiratory evaporation in the painted quail, *Excalfactoria chinensis*, during ontogeny of thermoregulation. Comp Biochem Physiol 38A:611-617

Bernstein MH (1971b) Cutaneus water loss in small birds. Condor 73:468-469

Berthold P (1975) Migration: Control and metabolic physiology. In: Farner DS, King JR (eds) Avian biology, vol V, Chap 2. Academic Press, London New York, pp 77-128

Bhattacharyya TK, Calas A, Assenmacher I (1975a) Adrenocortical response in the duck exposed to corticosteroid administration and salt loading. Cell Tissue Res 160:219-229

Bhattacharyya TK, Calas A, Assenmacher I (1975b) Effects of corticosteroid treatment and salt loading on the cytophysiology of the interrenal tissue in the pigeon and the quail. Gen Comp Endocrinol 26:115-125

Bielorai R, Harduf Z, Alumot E (1972) The free amino acid pattern of the intestinal contents of chicks fed raw and heated soybean meal. J Nutr 102:1377-1382

Bierer BW, Carll WI, Eleazer TH (1966) A comparison of the survival time and gross pathology in producing and non-producing white leghorn hens deprived of water. Poult Sci 45:65-67

Bindslev N (1979) Sodium transport in the hen lower intestine. Induction of sodium sites in the brush border by a low sodium diet. J Physiol (London) 288:449-466

Bindslev N, Skadhauge E (1971a) Salt and water permeability of the coprodeum and large intestine in the normal and dehydrated fowl (*Gallus domesticus*). In vivo perfusion studies. J Physiol (London) 216:735-751

176

Bindslev N, Skadhauge E (1971b) Sodium chloride absorption and solute-linked water flow across the epithelium of the corpodeum and large intestine in the normal and dehydrated fowl (*Gallus domesticus*). In vivo perfusion studies. J Physiol (London) 216:753–768

Boelkins JN, Mueller WJ, Hall KL (1973) Cardiac output distribution in the laying hen during shell formation. Comp Biochem Physiol 46A:735-743

Boissin J (1967) Le controle hypothalamo-hypophysaire de la fonction cortico-surrenalienne chez le canard. J Physiol (Paris) 59:423-444

Boissin J, Assenmacher I (1968) Rhytmes circadiens des taux sanguin et surrenalien de la corticosterone che la caille. C R Acad Sci Paris Ser D 267:2193-2196

Bokori J (1961) Neue Methode zur Harngewinnung von Geflügel. Acta Vet Acad Sci Hung 11:415-422

Bokori J, Boldizsár H, Kutas F (1965) On the physical properties and chemical composition of the urine of poultry. Acta Vet Acad Sci Hung 15:39-44

Bond CF, Gilbert PW (1958) Comparative study of blood volume in representative aquatic and nonaquatic birds. Am J Physiol 194:510-521

Boorman KN (1971) The renal reabsorption of arginine, lysine and ornithine in the young cockerel (*Gallus domesticus*). Comp Biochem Physiol 39A:29-38

Boorman KN, Falconer IR (1972) The renal reabsorption of histidine, leucine and the cationic amino acids in the young cockerel (*Gallus domesticus*). Comp Biochem Physiol 42A:311-320

Bouillé C, Baylé JD (1978) Comparison between hypothalamic multiple-unit activity of the hippocampus. Neuroendocrinology 25:303-309

Bouverot P, Hildwein G (1978) Combined effects of hypoxia and moderate heat load on ventilation in awake Pekin ducks. Respir Physiol 35:373-384

Bradley EL, Holmes WN (1971) The effects of hypophysectomy on adrenocortical function in the duck (*Anas platyrhynchos*). J Endocrinol 49:437-457

Bradley EL, Holmes WN, Wright A (1971) The effects of neurohypophysectomy on the pattern of renal excretion in the duck (*Anas platyrhynchos*). J Endocrinol 51: 57-65

Braun EJ (1976) Intrarenal blood flow distribution in the desert quail following salt loading. Am J Physiol 231:1111-1118

Braun EJ (1978) Renal response of the starling (*Sturnus vulgaris*) to an intravenous salt load. Am J Physiol 234:F270-F278

Braun EJ, Dantzler WH (1972) Function of mammalian-type and reptilian-type nephrons in kidney of desert quail. Am J Physiol 222:617-629

Braun EJ, Dantzler WH (1974) Effects of ADH on single-nephron glomerular filtration rates in the avian kidney. Am J Physiol 226:1-8

Braun EJ, Dantzler WH (1975) Effects of water load on renal glomerular and tubular function in desert quail. Am J Physiol 229:222-228

Bretz WL, Schmidt-Nielsen K (1971) Bird respiration: flow patterns in the duck lung. J Exp Biol 54:103-118

Brown Jr GW (1970) Nitrogen metabolism of birds. In: Campbell JW (ed) Comparative biochemistry of nitrogen metabolism. Academic Press, London New York pp 711-793

Brown KI, Brown DJ, Meyer RK (1958a) Effect of surgical trauma, ACTH and adrenal cortical hormones on electrolytes, water balance and gluconeogenesis in male chickens. Am J Physiol 192:43-50

Brown KI, Meyer RK, Brown DJ (1958b) A study of adrenalectomized male chickens with and without adrenal hormone treatment. Poult Sci 37:680-684

Brown WO, McCracken KJ (1965) The partition of certain mineral nutrients in the colostomised laying pullet and the determination of faecal endogenous calcium and phosphorus excretion by an dilution method. J Agric Sci 64:305-310

Browne TG (1922) Some observations on the digestive system of the fowl. J Comp Pathol 35:12-32

Burford HJ, Bond RF (1968) Avian cardiovascular parameters: effect of intravenous osmotic agents, relation to salt gland secretion. Experientia 24:1086-1088

Burgess WW, Harvey AM, Marshall EK (1933) The site of the antidiuretic action of pituitary extract. J Pharmacol Exp Ther 49:237-249

Burns CH, Cravens WW, Phillips PH (1952) The requirement of breeding hens for sodium chloride. Poult Sci 31:302-306

Burns CH, Cravens WW, Phillips PH (1953) The sodium and potassium requirements of the chick and their interrelationship. J Nutr 50:317-329

177

Butler DG (1980) Functional nasal salt glands in adrenalectomized domestic ducks (*Anas platyrhynchos*). Gen Comp Endocrinol 40:15-27

Cade TJ (1964) Water and salt balance in granivorous birds. In: Wayner MJ (ed) Thirst. Pergamon Press, Oxford, pp 237-256

Cade TJ, Bartholomew GA (1959) Sea-Water and salt utilization by savannah sparrows. Physiol Zool 32:230-238

Cade TJ, Dybas JA (1962) Water economy of the budgerygah. Auk 79:345-364

Cade TJ, Greenwald L (1966) Nasal salt secretion in falconiform birds. Condor 68:338-350

Cade TJ, MacLean GL (1967) Transport of water by adult sandgrouse to their young. Condor 69:323-343

Cade TJ, Tobin CA, Gold A (1965) Water economy and metabolism of two estrildine finches. Physiol Zool 38:9-33

Calder WA (1964) Gaseous metabolism and water relations of the zebra finch, *Taeniopygia castanotis*. Physiol Zool 37:400-413

Calder WA (1968) Nest sanitation: a possible factor in the water economy of the roadrunner. Condor 70:279

Calder WA, Bentley PJ (1967) Urine concentrations of two carnivorous birds, the white pelican and the roadrunner. Comp Biochem Physiol 22:607-609

Calder WA, Schmidt-Nielsen K (1967) Temperature regulation and evaporation in the pigeon and the roadrunner. Am J Physiol 213:883-889

Carey C, Morton ML (1971) A comparison of salt and water regulation in California quail (*Lophortyx californicus*) and Gambel's quail (*Lophortyx gambelii*). Comp Biochem Physiol 39A:75-101

Carpenter RE, Stafford MA (1970) The secretory rates and the chemical stimulus for secretion of the nasal salt glands in the Rallidae. Condor 72:316-324

Chan MY, Holmes WN (1971) Studies on a "renin-angiotensin" system in the normal and hypophysectomized pigeon (*Columba livia*). Gen Comp Endocrinol 16:304-311

Chapman TE, Black AL (1967) Water turnover in chickens. Poult Sci 46:761-765

Chapman TE, McFarland LZ (1971) Water turnover in coturnix quail with individual observations on a burrowing owl, petz conure and vulturine fish eagle. Comp Biochem Physiol 39A:653–656

Chapman TE, Mihai D (1972) Influences of sex and egg production on water turnover in chickens. Poult Sci 51:1252-1256

Chauvet J, Lenci M-T, Achér R (1960) Presence de deux vasopressines dans la neurohypophyse du poulet. Biochim Biophys Acta 38:571-573

Chavez E, Kratzer FH (1973) The potassium requirement of poults. Poult Sci 52:1542-1544

Choshniak I, Munck BG, Skadhauge E (1977) Sodium chloride transport across the chicken coprodeum. Basic characteristics and dependence on the sodium chloride intake. J Physiol (London) 271:489-504

Clara M (1926) Beiträge zur Kenntnis des Vogeldarmes. I. Teil. Mikroskopische Anatomie. Z Mikr Anat Forsch 4:346-416

Clark Jr GA (1979) Body weights of birds: A review. Condor 81:193-202

Clark NB, Wideman RF (1977) Renal excretion of phosphate and calcium in parathyroidectomized starlings. Am J Physiol 233:F138-F144

Clark NB, Braun EJ, Wideman RF (1976) Parathyroid hormone and renal excretion of phosphate and calcium in normal starlings. Am J Physiol 231:1152-1158

Clemens ET, Stevens CE, Southworth M (1975) Sites of organic acid production and pattern of digesta movement in the gastrointestinal tract of geese. J Nutr 105:1341-1350

Cloudsley-Thompson JL, Mohamed ERM (1967) Water economy of the ostrich. Nature (London) 216:1040

Cohen I, Hurwitz S, Bar A (1971) Acid-base balance and sodium-to-chloride ratio in diets of laying hens. J Nutr 102:1-8

Cohen RR (1967a) Anticoagulation, centrifugation time, and sample replicate number in the microhematocrit method for avian blood. Poult Sci 46:214-218

Cohen RR (1967b) Total circulating erythrocyte and plasma volumes of ducks measured simultaneously with Cr^{51} Poult Sci 46:1539-1544

Colvin LB, Creger CR, Couch JR, Ferguson TM, Ansari MNA (1966) A simplified method for separation of urine and feces in the immature fowl. Proc Soc Exp Biol Med 123:415-417

178

Cooch FG (1964) A preliminary study of the survival value of a functional salt gland in prairie Anatidae. Auk 81:380-393

Cooke HJ, Young JA (1970) Development of glomerular filtration rate and electrolyte and osmolal clearance in the late-embryonic and newly hatched chicken. Pflügers Arch 318:315-324

Coon JM (1939) A new method for the assay of posterior pituitary extracts. Arch Int Pharmacodyn 62:79-99

Coulson EJ, Hughes JS (1930) Collection and analysis of chicken urine. Poult Sci 10:53-57

Crane RK (1977) The gradient hypothesis and other models of carrier-mediated active transport. Rev Physiol Biochem Pharmacol 78:99-159

Crawford EC, Lasiewski RC (1968) Oxygen consumption and respiratory evaporation of the emu and rhea. Condor 70:333-339

Crawford EC, Schmidt-Nielsen K (1967) Temperature regulation and evaporative cooling in the ostrich. Am J Physiol 212:347-353

Crocker AD, Holmes WN (1971) Intestinal absorption in ducklings (*Anas platyrhynchos*) maintained on fresh water and hypertonic saline. Comp Biochem Physiol 40A: 203-211

Crocker AD, Cronshaw J, Holmes WN (1975) The effect of several crude oils and some petroleum distillation fractions on intestinal absorption in ducklings (*Anas platyrhynchos*). Environ Physiol Biochem 5:92-106

Crompton DWT (1966) Measurements of glucose and amino acid concentrations, temperature and pH in the habitat of polymorphus minutus (Acanthocephala) in the intestine of domestic ducks. J Exp Biol 45:279-284

Crompton DWT, Edmonds SJ (1969) Measurements of the osmotic pressure in the habitat of polymorphus minutus (Acanthocephala) in the intestine of domestic ducks. J Exp Biol 50:69-77

Crompton DWT, Nesheim MC (1970) Lipid, bile acid, water and dry matter content of the intestinal tract of domestic ducks with reference to the habitat of polymorphus minutus (Acanthocephala). J Exp Biol 52:437-445

Culbert J, Wells JW (1975) Aspects of adrenal function in the domestic fowl. J Endocrinol 65:363-376

Dantzler WH (1966) Renal response of chickens to infusion of hyperosmotic sodium chloride solution. Am J Physiol 210:640-646

Dantzler WH (1970) Kidney function in desert vertebrates. In: Benzon GK, Phillips JG (eds) Memoirs of the Society for Endocrinology. Horm Environ 18:157-190

Dantzler WH (1978) Urate excretion in nonmammalian vertebrates. In: Kelley WN, Weiner IM (eds) Uric Acid. Chap 8. Springer, Berlin Heidelberg New York, pp 185-210

Davies SJJF (1977) The timing of breeding by the zebra finch *Taeniopygia castanotis* at Mileura, Western Australia. Ibis 119:369-372

Davis RE (1927) The nitrogenous constituents of hen urine. J Biol Chem 74:509-513

Dawson WR (1958) Relation of oxygen consumption and evaporative water loss to temperature in the cardinal. Physiol Zool 31:37-48

Dawson WR (1965) Evaporative water losses of some australian parrots. Auk 82:106-108

Dawson WR, Bartholomew GA (1968) Temperature regulation and water economy of desert birds. In: Brown GW (ed) Desert biology. Special topics on the physical and biological aspects of arid regions. Academic Press, London New York pp 357-394

Dawson WR, Bennett AF (1973) Roles of metabolic level and temperature regulation in the adjustment of western plumed pigeons (*Lophophaps ferruginea*) to desert conditions. Comp Biochem Physiol 44A: 249-266

Dawson WR, Fisher CD (1969) Responses to temperature by the spotted nightjar (*Eurostopodus guttatus*). Condor 71:49-53

Dawson WR, Hudson JW (1970) Birds. In: Whittow GC (ed) Comparative physiology of thermoregulation, vol I. Academic Press, London New York, pp 223-310

Dawson WR, Schmidt-Nielsen K (1964) Terrestrial animals in dry heat: desert birds. In: Dill DB, Adolph EF, Wilber CG (eds) Handbook of physiology, Section 4, Chap 31. American Physiological Society, Washington DC, pp 481-492

Dawson WR, Shoemaker VH, Tordoff HB, Borut A (1965) Observations on the metabolism of sodium chloride in the red crossbill. Auk 82:606-623

Dawson WR, Bennett AF, Hudson JW (1978) Metabolism and thermoregulation in hatchling ring-billed gulls. Condor 78:49-60

Dawson WR, Carey C, Adkisson CS, Ohmart RD (1979) Responses of Brewer's and chipping sparrows to water restriction. Physiol Zool 52:529

Desmedt JE, Delwaide PJ (1966) Physiological experimentation on the pigeon. Lab Anim Care 16:191-197

Deutsch H, Hammel HT, Simon E, Simon-Oppermann C (1979) Osmolality and volume factors in salt gland control of pekin ducks after adaptation to chronic salt loading. J Comp Physiol 129:301-308

Dicker SE, Haslam J (1966) Water diuresis in the domestic fowl. J Physiol (London) 183:225-235

Dicker SE, Haslam J (1972) Effects of exteriorization of the ureters on the water metabolism of the domestic fowl. J Physiol (London) 224:515-520

Dicker SE, Eggleton MG, Haslam J (1966) The effects of urea and hydrochlorothiazide on the renal functions of rat and domestic fowl. J Physiol (London) 187:247-255

Diehl B, Kurowski C, Myrcha A (1972) Changes in the gross chemical composition and energy content of nestling red-backed shrikes (*Lanius collurio L.*). Bull Acad Pol Sci 20:837-843

Dixon JM (1958) Investigation of urinary water reabsorption in the cloaca and rectum of the hen. Poult Sci 37:410-414

Dixon JM, Wilkinson WS (1957) Surgical technique for exteriorization of the ureters of the chicken. Am J Vet Res 18:665-667

Donaldson EM. Holmes WN (1965) Corticosteroidogenesis in the freshwater and saline maintained duck (*Anas platyrhynchos*). Endocrinology 32:329-336

Douglas DS (1968) Salt and water metabolism of the Adelie penguin. Antarct Res Ser 12:167-190

Douglas DS (1970) Electrolyte excretion in seawater-loaded herring gulls. Am J Physiol 219:534-539

Duke GE, Petrides GA, Ringer RK (1968) Chromium-51 in food metabolizability and passage rate studies with the ring-necked pheasant. Poult Sci 47:1356-1364

Duke GE, Dziuk HE, Evanson OA (1969) Fluxes of ions, glucose, and water in isolated jejunal segments in normal and bluecomb diseased turkeys. Poult Sci 48:2114-2123

Duke GE, Ciganek JG, Evanson OA (1973) Food consumption and energy, water, and nitrogen budgets in captive great-horned (*Bubo virginianus*). Comp Biochem Physiol 44A:283-292

Duke GE, Kostuch TE, Evanson OA (1975) Electrical activity and intraluminal pressures in the lower small intestine of turkeys. Am J Dig Dis 20:1040-1046

Duncan CJ (1962) Salt preferences of birds and mammals. Physiol Zool 35:120-132

Dunson WA (1970) Excessive drinking (Polydipsia) in a galapagos mockingbird (*Nesomimus*). Comp Biochem Physiol 36:143-151

Dunson WA, Buss EG (1968) Abnormal water balance in a mutant strain of chickens. Science 161:167-169

Dunson WA, Buss EG, Sawyer WH, Sokol HW (1972) Hereditary polydipsia and polyuria in chickens. Am J Physiol 222:1167-1176

Dunson WA, Dunson MK, Ohmart RD (1976) Evidence for the presence of nasal salt glands in the roadrunner and the coturnix quail. Exp Zool 198:209-216

Dusseau JW, Meier AH (1971) Diurnal and seasonal variations of plasma adrenal steroid hormone in the white-throated sparrow, *Zonotrichia albicollis*. Gen Comp Endocrinol 16:399-408

Edwards WH, Wilson WO (1954) Relationship of hyperthermy to nitrogen excretion in chickens. Am J Physiol 179:76-78

Ellis RA, Goertemiller CC, DeLellis RA, Kablotsky YH (1963) The effect of a salt and water regimen on the development of the salt glands of domestic ducklings. Dev Biol 8:286-308

Ellis RA, Goertemiller CC, Stetson DL (1977) Significance of extensive 'leaky' cell junctions in the avian salt gland. Nature (London) 268:555-556

Emery N, Poulson TL, Kinter WB (1972) Production of concentrated urine by avian kidneys. Am J Physiol 223:180-187

Ensor DM (1975) Prolactin and adaptation. Symp Zool Soc London 35:129-148

Ensor DM, Phillips JG (1970) The effect of salt loading on the pituitary prolactin levels of the domestic duck (*Anas platyrhynchos*) and juvenile herring or lesser black-backed gulls (*Larus argentatus* or *Larus fuscus*). J Endocrinol 48:167-172

Ensor DM, Phillips JG (1972a) The effect of age and environment on extrarenal salt excretion in juvenile gulls (*Larus argentatus* and *L. fuscus*). J Zool London 168:119-126

Ensor DM, Phillips JG (1972b) The effect of dehydration on salt and water balance in gulls (*Larus argentatus* and *L. fuscus*). J Zool London 168:127-137

Erasmus T (1978a) The handling of constant volumes of various concentrations of seawater by the jackass penguin *Spheniscus demersus*. Zool Afr 13:71-80

Erasmus T (1978b) The relative importance of the various electrolyte excretory pathways in osmotically stressed penguins. Comp Biochem Physiol 59A:379-384

Ernst SA, Ellis RA (1969) The development of surface specialization in the secretory epithelium of the avian salt gland in response to osmotic stress. J Cell Biol 40:305-321

Ernst SA, Mills JW (1977) Basolateral plasma membrane localization of ouabain-sensitive sodium transport sites in the secretory epithelium of the avian salt gland. J Cell Biol 75:74-94

Fänge R, Schmidt-Nielsen K, Osaki H (1958a) The salt gland of the herring gull. Biol Bull 115: 162-171

Fänge R, Schmidt-Nielsen K, Robinson M (1958b) Control of secretion from the avian salt gland. Am J Physiol 195:321-326

Farland DJ, Baker E (1968) Factors affecting feather posture in the barbary dove. Anim Behav 16:171-177

Farner DS, King JR (1972-75) Avian biology. vol 1-5. Academic Press London New York

Farner DS, Oksche A (1962) Neurosecretion in birds. Gen Comp Endocrinol 2:113-147

Feldotto A (1929) Die Harnkanälchen des Huhnes. Z Mikrosk Anat Forsch 17:353-370

Fellows GJ, Turnbull GJ (1971) The permeability of mammalian urinary bladder epithelium. Eur J Clin Biol Res 16:303-310

Fenna L, Boag DA (1974a) Filling and emptying of the galliform caecum. Can J Zool 52:537-540

Fenna L, Boag DA (1974b) Adaptive significance of the caeca in Japanese quail and spruce grouse (Galliformes). Can J Zool 52:1577-1584

Fisher CD, Lindgren E, Dawson WR (1972) Drinking patterns and behavior of Australian desert birds in relation to their ecology and abundance. Condor 74:111-136

Fitzsimons JT (1978) The role of the renin-angiotensin system in the regulation of extracellular fluid volume. In: Jørgensen CB, Skadhauge E (eds) Osmotic and volume regulation. Munksgaard, Copenhagen, pp 100-115

Fletcher GL, Holmes WN (1968) Observations on the intake of water and electrolytes by the duck (*Anas platyrhynchos*) maintained on fresh water and on hypertonic saline. J Exp Biol 49:325-339

Fletcher GL, Stainer IM, Holmes WN (1967) Sequential changes in the adenosinetriphosphatase activity and electrolyte excretory capacity of the nasal glands of the duck (*Anas platyrhynchos*), during the period of adaptation to hypertonic saline. J Exp Biol 47:375-392

Fogden MPL (1972) Premigratory dehydration in the reed warbler *Acrocephalus scirpaceus* and water as a factor limiting migratory range. Ibis 114:548-552

Folk RL (1969) Spherical urine in birds: Petrografy. Science 166:1516-1518

Follett BK, Farner DS (1966) The effects of the daily photoperiod on gonadal growth, neurohypophysial hormone content, and neurosecretion in the hypothalamo-hypophysial system of the Japanese quail (*Coturnix coturnix japonica*). Gen Comp Endocrinol 7:111-124

Freeman BM (1971) Extravascular fluids. In: Bell DJ, Freeman BM (eds) Physiology and biochemistry of the domestic fowl, vol II, Chap 44. Academic Press, London New York, pp 981-984

Freeman BM, Manning ACC (1977) Factors affecting the responses of Gallus domesticus to glucagon, adrenocorticotrophic hormone and theophylline. Comp Biochem Physiol 57A:215-219

Fussell MH (1960) Collection of urine and faeces from the chicken. Nature (London) 185:332-333

Gadow H (1879a) Versuch einer vergleichenden Anatomie des Verdauungssystems der Vögel. I Teil. Jena Z Med Naturwiss 13:92-171

Gadow H (1879b) Ibid. II Teil 13:339-403

Gandal CP (1969) Avian anesthesia. Fed Proc 28:1533-1534

Gasaway WC (1976a) Volatile fatty acids and metabolizable energy derived from cecal fermentation. Comp Biochem Physiol 53A:115-121

Gasaway WC (1976b) Seasonal variation in diet, volatile fatty acid production and size of the cecum of rock ptarmigan. Comp Biochem Physiol 53A:109-114

Gasaway WC, Holleman DF, White RG (1975) Flow of digesta in the intestine and cecum of the rock ptarmigan. Condor 77:467-474

Gasaway WC, White RG, Holleman DF (1976) Digestion of dry matter and absorption of water in the intestine and cecum of rock ptarmigan. Condor 78:77-84

George U (1969) Über das Trinken der Jungen und andere Lebensäußerungen des Senegal-Flughuhns, *Pterocles senegallus* in Marokko. J Ornithol 110:181–191

Gibbs OS (1929) The secretion of uric acid by the fowl. Am J Physiol 88:87-100

Gilbert AB (1961) The innervation of the renal portal valve of the domestic fowl. J Anat 95:594-598

Gill JB, Burford HJ (1967) Secretion from normal and supersensitive avian salt glands. J Exp Zool 168:451-454

Gilmore JP, Dietz J, Gilmore C, Zucker IH (1977) Evidence for a chloride pump in the salt gland of the goose. Comp Biochem Physiol 56A:121-126

Graber JW, Nalbandov AV (1965) Neurosecretion in the white leghorn cockerel. Gen Comp Endocrinol 5:485-492

Grabowski CT (1967) Ontogenetic changes in the osmotic pressure and sodium and potassium concentrations of chick embryo serum. Comp Biochem Physiol 21:345-350

Greenwald L, Stone WB, Cade TJ (1967) Physiological adjustments of the budgerygah (*Melopsittacus undulatus*) to dehydrating conditions. Comp Biochem Physiol 22:91-100

Greschik E (1912) Mikroskopische Anatomie des Enddarms der Vögel. Aquila 19:210-269

Haas W, Beck P (1979) Zum Frühjahrszug palaarktischer Vögel über die westliche Sahara. J Ornithol 120:237-246

Halnan ET (1949) The architecture of the avian gut and tolerance of crude fibre. Br J Nutr 3:245-253

Halpin JG, Holmes CE, Hart EB (1936) Salt requirements of poultry. Poult Sci 15:99-103

Hanwell A, Peaker M (1975) The control of adaptive hypertrophy in the salt glands of geese and ducks. J Physiol (London) 248:193-205

Hanwell A, Linzell JL, Peaker M (1971a) Salt-gland secretion and blood flow in the goose. J Physiol (London) 213:373-387

Hanwell A, Linzell JL, Peaker M (1971b) Cardiovascular responses to salt-loading in conscious domestic geese. J Physiol (London) 213:389-398

Hanwell A, Linzell JL, Peaker M (1972) Nature and location of the receptors for salt-gland secretion in the goose. J Physiol (London) 226:453-472

Harriman AE (1967) Laughing gulls offered saline in preference and survival tests. Physiol Zool 40:273-279

Harriman AE, Kare MR (1966a) Tolerance for hypertonic saline solutions in herring gulls, starlings, and purple grackles. Physiol Zool 39:117-122

Harriman AE, Kare MR (1966b) Aversion to saline solutions in starlings, purple grackles, and herring gulls. Physiol Zool 39:123-126

Harris KM, Koike TI (1977) The effects of dietary sodium restriction on fluid and electrolyte metabolism the chicken (*Gallus domesticus*). Comp Biochem Physiol 58A:311-317

Hart JS, Berger M (1972) Energetics, water economy and temperature regulation during flight. In: Voous KH (ed) Proc XVth Int Ornithol Congr. EJ Brill, Leiden, pp 189-199

Hart JS, Roy OZ (1966) Respiratory and cardiac responses to flight in pigeons. Physiol Zool 39:291-306

Hart WM, Essex HE (1942) Water metabolism of the chicken (*Gallus domesticus*) with special reference to the role of the cloaca. Am J Physiol 136:657-668

Hartman FA (1946) Adrenal and thyroid weights in birds. Auk 63:42-64

Hasan SH, Heller H (1968) The clearance of neurohypophysial hormones from the circulation of non-mammalian vertebrates. Br J Pharmacol Chemother 33:523-530

Heller H, Hasan SH (1968) The fate of vasotocin and oxytocin in the domestic fowl. Arch Anat Histol Embryol 51:315-319

Heller H, Pickering BT (1961) Neurohypophysial hormones of non-mammalian vertebrates. J Physiol (London) 155:98-114

Helton ED, Holmes WN (1973) The distribution and metabolism of labelled corticosteroids in the duck (*Anas platyrhynchos*). J Endocrinol 56:361-385

Hesse W, Lustick S (1977) A comparison of the water requirements of marsh and upland redwing blackbirds (*Agelaius phoeniceus*). Physiol Zool 45:196-203

Hester HR, Essex HE, Mann FC (1940) Secretion of urine in the chicken (*Gallus domesticus*). Am J Physiol 128:592-602

Heuser GF (1952) Salt additions to chick rations. Poult Sci 31:85-88

Hirano T (1964) Further studies on the neurohypophysial hormones in the avian median eminence. Endocrinol Jpn 11:87-95

Hokin MR (1967) The Na, K, and Cl content of goose salt gland slices and the effects of acetylcholine and ouabain. J Gen Physiol 50:2197-2209

182

Holmes WN (1965) Some aspects of osmoregulation in reptiles and birds. Arch Anat Micr Morphol Exp 54:491-514

Holmes WN (1972) Regulation of electrolyte balance in marine birds with special reference to the role of the pituitary-adrenal axis in the duck (*Anas platyrhynchos*). Fed Proc 31:1587-1598

Holmes WN (1975) Hormones and osmoregulation in marine birds. Gen Comp Endocrinol 25:249-258

Holmes WN (1978) Endocrine response in osmoregulation: hormonal adaptation in aquatic birds. In: Assenmacher I, Farner DS (eds) Environmental endocrinology. Springer, Berlin Heidelberg New York, pp 230-239

Holmes WN, Adams BM (1963) Effects of adrenocortical and neurohypophysial hormones on the renal excretory pattern in the water-loaded duck (*Anas platyrhynchos*). Endocrinology 73:5-10

Holmes WN, Phillips JG (1976) The adrenal cortex of birds In: Chester Jones I, Henderson IW (eds) General, comparative and clinical endocrinology of the adrenal cortex. Academic Press, London New York, pp 293-420

Holmes WN, Butler DG, Phillips JG (1961a) Observations on the effect of maintaining glaucous-winged gulls (*Larus glaucescens*) on fresh water and sea water for long periods. J Endocrinol 23:53-61

Holmes WN, Phillips JG, Butler DG (1961b) The effect of adrenocortical steroids on the renal and extra-renal response of the domestic duck (*Anas platyrhynchos*) after hypertonic saline loading. Endocrinology 69:483-495

Holmes WN, Fletcher GL, Stewart DJ (1968) The patterns of renal electrolyte excretion in the duck (*Anas platyrhynchos*) maintained on freshwater and on hypertonic saline. J Exp Biol 48:487-508

Holmes WN, Lockwood LN, Bradley EL (1972) Adenohypophysial control of extrarenal excretion in the duck (*Anas platyrhynchos*). Gen Comp Endocrinol 18:59-68

Holmes WN, Broock RL, Devlin JM (1974) Tritiated corticosteroid metabolism in intact and adenohypophysectomized ducks (*Anas platyrhynchos*). Gen Comp Endocrinol 22:417-427

Hossler FE, Sarras MP, Barrnett RJ (1978) Ouabain binding during plasma membrane biogenesis in duck salt gland. J Cell Sci 31:179-197

Howard BR (1975) Water balance of the hen during egg formation. Poult Sci 54:1046-1053

Huber GC (1917) On the morphology of the renal tubulus of vertebrates. Anat Res 13:305-339

Hudson DA, Levin RJ, Smyth DH (1971) Absorption from the alimentary tract. In: Bell DJ, Freeman BM (eds) Physiology and biochemistry of the domestic fowl, vol I, Chap 3. Academic Press, London New York, pp 51-71

Hudson JW, Dawson WR, Hill RW (1974) Growth and development of temperature regulation in nestling cattle egrets. Comp Biochem Physiol 49A:717-741

Hughes MR (1968) Renal and extrarenal sodium excretion in the common tern *Sterna hirundo*. Physiol Zool 41:210-219

Hughes MR (1969) Ionic and osmotic concentration of tears of the gull, *Larus glaucescens*. Can J Zool 47:1337-1339

Hughes MR (1970a) Cloacal and salt-gland ion excretion in the seagull, *Larus glaucescens*, acclimated to increasing concentrations of sea water. Comp Biochem Physiol 32:315-325

Hughes MR (1970b) Flow rate and cation concentration in salt gland secretion of the glaucous-winged gull, *Larus glaucescens*. Comp Biochem Physiol 32:807-812

Hughes MR (1970c) Relative kidney size in nonpasserine birds with functional salt glands. Condor 72:164-168

Hughes MR (1970d) Some observations on ion and water balance in the puffin, *Fratercula arctica*. Can J Zool 48:479-482

Hughes MR (1972a) Hypertonic salt gland secretion in the glaucous-winged gull, *Larus glaucescens*, in response to stomach loading with dilute sodium chloride. Comp Biochem Physiol 41A:121-127

Hughes MR (1972b) The effect of salt gland removal on cloacal ion and water excretion in the growing kittiwake, *Rissa tridactyla*. Can J Zool 50:603-610

Hughes MR (1975) Salt gland secretion produced by the gull, *Larus glaucescens*, in response to stomach loads of different sodium and potassium concentration. Comp Biochem Physiol 51A:909-913

Hughes MR (1976a) The effects of salt water adaptation on the Australian black swan, *Cygnus atratus* (Latham). Comp Biochem Physiol 55:271-277

Hughes MR (1976b) Effect of glucose on salt gland secretion in the glaucous-winged gull, *Larus glaucescens*. Comp Biochem Physiol 53A:311-312

Hughes MR (1977) Observations on osmoregulation in glaucous-winged gulls, *Larus glaucescens*, following removal of the supraorbital salt glands. Comp Biochem Physiol 57A:281-287

Hughes MR, Blackman JG (1973) Cation content of salt gland secretion and tears in the brolga, *Grus rubicundus* (Perry) (Aves: Gruidae). Aust J Zool 21:515-518

Hughes MR, Ruch FE (1969) Sodium and potassium in spontaneously produced salt gland secretion and tears of ducks, *Anas platyrhynchos*, acclimated to fresh and saline waters. Can J Zool 47:1133-1138

Hughes BO, Whitehead CC (1974) Sodium deprivation, feather pecking and activity in laying hens. Br Poult Sci 15:435-439

Hunsaker WG (1968) Blood volume of geese treated with androgen and estrogen. Poult Sci 47:371-376

Hunt JR, Hunsaker WG, Aitken JR (1964) Physiology of the growing and adult goose. Br Poult Sci 5-6:257-262

Hurwitz S, Bar A (1969) Relation between the lumen-blood electrochemical potential difference of calcium, calcium absorption and calcium-binding protein in the intestine of the fowl. J Nutr 99:217-224

Hurwitz S, Bar A, Clarkson TW (1970) Intestinal absorption of sodium and potassium in the laying fowl. J Nutr 100:1181-1187

Hurwitz S, Cohen I, Bar A, Bornstein S (1973) Sodium and chloride requirements of the chick: relationship to acid-base balance. Poult Sci 52:903-909

Hutchinson JCD (1955) Evaporative cooling in fowls. J Agric Sci 45:48-59

Hydén S, Knutson P-G (1959) Renal clearance and distribution volume of polyethylene glycol and inulin in the chicken. K Lantbrhoegsk Annlr 25:253-259

Imabayashi K, Kametaka M, Hatano T (1955) Studies on the digestion in the domestic fowl I. "Artificial anus operation" for the domestic fowl and the passage of the indicator throughout the digestive tract. Tohoku J Agric Res 6:99-117

Inman DL (1973) Cellulose digestion in ruffed grouse, Chukar Partridge, and Bobwhite quail. J Wildl Manage 37:114-121

Inman DL, Ringer RK (1973) Differential passage of cellulose and $^{51}CrCl_3$ in three non-domestic gallinaceous species. Poult Sci 52:1399-1405

Ishii S, Hirano T, Kobayashi H (1962) Neurohypophysial hormones in the avian median eminence and pars nervosa. Gen Comp Endocrinol 2:433-440

John TM, George JC (1977) Blood levels of cyclic AMP, thyroxine, uric acid, certain metabolites and electrolytes under heat-stress and dehydration in the pigeon. Arch Int Physiol Biochim 85:571-582

Johnson OW (1968) Some morphological features of avian kidneys. Auk 85:216-228

Johnson OW (1974) Relative thickness of the renal medulla in birds. J Morphol 142:277-284

Johnson OW, Mugass JN (1970a) Some histological features of avian kidneys. Am J Anat 127:423-436

Johnson OW, Mugaas JN (1970b) Quantitative and organizational features of the avian renal medulla. Condor 72:288-292

Johnson OW, Ohmart RD (1973a) Some features of water economy and kidney microstructure in the large-billed savannah sparrow (*Passerculus sandwichensis rostratus*). Physiol Zool 46:276-284

Johnson OW, Ohmart RD (1973b) The renal medulla and water economy in vesper sparrows (*Pooecetes gramineus*). Comp Biochem Physiol 44A:655-661

Johnson OW, Skadhauge E (1975) Structural-functional correlations in the kidneys and observations of colon and cloaca morphology in certain Australian birds. J Anat 120:495-505

Johnson OW, Phipps GL, Mugaas JN (1972) Injection studies of cortical and medullary organization in the avian kidney. J Morphol 136:181-190

Johnston DW (1968) Body characteristics of palm warblers following an overwater flight. Auk 85:13-18

Jones DR, Johansen K (1972) The blood vascular system of birds. In: Farner DS, King JR (eds) Avian biology, vol II, Chap 4. Academic Press, London New York, pp 157-285

Joubert CSW, MacLean GL (1973) The structure of the water-holding feathers of the Namaqua sandgrouse. Zool Africana 8:141-152

Katayama T (1924) Über die Verdaulichkeit der Futtermittel bei Hühnern. Bull Imp Agric Exp Stn Jpn 3:1-78

184

Katsoyannis PG, Vigneaud V du (1958) Arginine-vasotocin, a synthetic analogue of the posterior pituitary hormones containing the ring of oxytocin and the side-chain of vasopressin. J Biol Chem 233:1352-1354

Kaul R, Hammel HT (1979) Dehydration elevates osmotic threshold for salt gland secretion in the duck. Am J Physiol 237:R355-R359

Kellerup SU, Parker JE, Arscott GH (1965) Effect of restricted water consumption on broiler chickens. Poult Sci 44:78-83

King JR, Farner DS (1964) Terrestrial animals in humid heat: birds. In: Dill DB, Adolph EF, Wilber CG (eds) Handbook of physiology, Sect 4, Chap 38. American Physiological Society, Washington DC, pp 603-624

Kitchell RL, Strom L, Zotterman Y (1959) Electrophysiological studies of thermal and taste reception in chickens and pigeons. Acta Physiol Scand 46:133-151

Kobayashi H (1978) Evolution of the target organ. In: Gaillard PJ, Boer HH (eds) Comparative endocrinology. Elsevier/North Holland, Amsterdam, pp 401-404

Koike TI, McFarland LZ (1966) Urography in the unanesthetized hydropenic chicken. Am J Vet Res 27:1130-1133

Koike T, Lepkovsky S (1967) Hypothalamic lesions producing polyuria in chickens. Gen Comp Endocrinol 8:397-402

Koike TI, Pryor LR, Neldon HL, Venable RS (1977) Effect of water deprivation on plasma radioimmunoassayable arginine vasotocin in conscious chickens. Gen Comp Endocrinol 33:359-364

Koike TI, Pryor LR, Neldon HL (1979) Effect of saline infusion on plasma immunoreactive vasotocin in conscious chickens. Gen Comp Endocrinol 37:451-458

Korr IM (1939) The osmotic function of the chicken kidney. J Cell Comp Physiol 13:175-194

Krag B, Skadhauge E (1972) Renal salt and water excretion in the budgerygah (*Melopsittacus undulatus*). Comp Biochem Physiol 41A:667-683

Krahower CA, Heino HE (1947) Relationship of growth and nutrition to cardiorenal changes induced in birds by a high salt intake. Arch Pathol 44:143-162

Kripalani K, Ghosh A, Rahman H (1967) Hypothalamic neurosecretion in relation to water deprivation in ploceids of arid and swampy zones. J Morphol 123:35-42

Krista LM, Carlson CW, Olson OE (1961) Some effects of saline waters on chicks, laying hens, poults, and ducklings. Poult Sci 40:938-944

Krista LM, Lisano ME, McDaniel GR, Mora EC (1979) Plasma corticosterone values in genetic hypertensive and hypotensive strain of turkeys. Poult Sci 58:252-254

Krogh A (1939) Osmotic regulation in aquatic animals, Unabridged republication by Dover Publications, New York (1965), Cambridge University Press, pp 1-242

Kubicek JJ, Sullivan TW (1973) Dietary chloride requirement of starting turkeys. Poult Sci 52:1903-1909

Kuenzel WJ, Helms CW (1970) Hyperphagia, polydipsia, and other effects of hypothalamic lesions in the white-throated sparrow, *Zonotrichia albicollis*. Condor 72:66-75

Kumpost HE, Sullivan TW (1966) Minimum sodium requirement and interaction of potassium and sodium in the diet of young turkeys. Poult Sci 45:1334-1339

Lai HC, Duke GE (1978) Colonic motility in domestic turkeys. Am J Dig Dis 23:673-681

Lambert PP (1945) Etude comparé de l'elimination de l'inuline, de la creatinine et de L'Urosélectan B par le rein des oiseaux. Arch Int Pharmacodyn 71:313-342

Lange R, Staaland H (1966) Anatomy and physiology of the salt gland in the grey heron, *Ardea cinerea*. Nytt Mag Zool 13:5-9

Langford HG, Fallis N (1966) Diuretic effect of angiotensin in the chicken. Proc Soc Exp Biol Med 123:317-321

Lanthier A, Sandor T (1967) Control of the salt-secreting gland of the duck. Can J Physiol Pharmacol 45:925-936

Lanthier A, Sandor T (1973) The effect of 18-hydroxycorticosterone on the salt-secreting gland of the duck (*Anas platyrhynchos*). I. Osmotic regulation. Can J Physiol Pharmacol 51:776-778

Lasiewski RC (1972) Respiratory functions in birds. In: Farner DS, King JR (eds) Avian biology vol II Chap 5. Academic Press, London New York, pp 287-342

Lasiewski RC, Bartholomew GA (1966) Evaporative cooling in the poorwill and the tawny frogmouth. Condor 68:253-262

Lasiewski RC, Bernstein MH (1971) Cutaneus water loss in the roadrunner and poor-will. Condor 73:470-472

Lasiewski RC, Dawson WR (1967) A re-examination of the relation between standard metabolic rate and body weight in birds. Condor 69:13-23

Lasiewski RC, Snyder GK (1969) Responses to high temperature in nestling double-crested and pelagic cormorants. Auk 86:529-540

Lasiewski RC, Acosta ML, Bernstein MH (1966a) Evaporative water loss in birds-I. Characteristics of the open flow method of determination, and their relation to estimates of thermoregulatory ability. Comp Biochem Physiol 19:445-457

Lasiewski RC, Acosta AL, Bernstein MH (1966b) Evaporative water loss in birds-II. A modified method for determination by direct weighing. Comp Biochem Physiol 19:459-470

Lasiewski RC, Bernstein MH, Ohmart RD (1971) Cutaneous water loss in the roadrunner and poor will. Condor 73:470-472

Lauson HD (1967) Metabolism of antidiuretic hormones. Am J Med 42:713-744

Lawzewitsch I von, Sarrat R (1970) Das neurosekretorische Zwischenhirn-Hypophysensystem von Vögeln nach langer osmotischer Belastung. Acta Anat 77:521-539

Leach RM, Nesheim MC (1963) Studies on chloride deficiency in chicks. J Nutr 81:193-199

Lee P, Schmidt-Nielsen K (1971) Respiratory and cutaneous evaporation in the zebra finch: effect on water balance. Am J Physiol 220:1598-1605

LeFebvre EA (1964) The use of D_2O for measuring energy metabolism in Columba livia at rest and in flight. Auk 81:403–416

Legait H, Legait E (1955) Nouvelles recherches sur les modifications de structure de lobe distal de l'hypophyse au cours de divers états physiologiques et experimentaux chez la poule Rhode-Island. C R Assoc Anat 91:902-907

Leopold AS (1953) Intestinal morphology of gallinaceous birds in relation to food habits. J Wildl Managem 17:197-203

Lifschitz E, German O, Favret EA, Manso F (1967) Difference in water ingestion associated with sex in poultry. Poult Sci 46:1021-1023

Ligon DJ (1969) Some aspects of temperature relations in small owls. Auk 86:458-472

Lind J, Munck BG, Olsen O (1980a) Effects of dietary intake of sodium chloride on sugar and amino acid transport across isolated hen colon. J Physiol (London) 305:327-336

Lind J, Munck BG, Olsen O, Skadhauge E (1980b) Effects of sugars, amino acids and inhibitors on electrolyte transport across hen colon at different sodium chloride intakes. J Physiol (London) 305:315-325

Long S, Skadhauge E (1980) Renal reabsorption of Na and K in *Gallus*: Role of urinary precipitates. Acta Physiol Scand 109:31A

Lonsdale K, Sutor DJ (1971) Uric acid dihydrate in bird urine. Science 172:958-959

Lopez GA, Phillips RW, Nockels CF (1973) The effect of age on water metabolism in hens. Proc Soc Exp Biol Med 143:545-547

Louw GN (1972) The role of advective fog in the water economy of certain Namib desert animals. Symp Zool Soc Lond 31:297-314

Louw GN, Belonje PC, Coetzee HJ (1969) Renal function, respiration, heart rate and thermoregulation in the ostrich (*Struthio camelus*). Sci Pap Namib Desert Res Stn 42:43-54

Lumijarvi DH, Vohra P (1976) Studies on the sodium requirement of growing japanese quail. Poult Sci 55:1410-1414

Lustick S (1970) Energetics and water regulation in the cowbird (*Molothrus ater obscurus*). Physiol Zool 43:270-287

Lyngdorf-Henriksen P, Munck BG, Skadhauge E (1978) Sodium chloride transport across the lower intestine of the chicken. Dependence on sodium chloride concentration and effect of inhibitors. Pflügers Arch Ges Physiol 378:161-165

Macchi IA, Phillips JG, Brown P (1967) Relationship between the concentration of corticosteroids in avian plasma and nasal gland function. J Endocrinol 38:319-329

MacMillen RE (1962) The minimum water requirements of mourning doves. Condor 64:165-166

MacMillen RE, Trost CH (1966) Water economy and salt balance in white-winged and inca doves. Auk 83:441-456

MacMillen RE, Trost Ch (1967) Thermoregulation and water loss in the inca dove. Comp Biochem Physiol 20:263-273

186

MacMillen RE, Whittow GC, Christopher EA, Ebisu RJ (1977) Oxygen consumption, evaporative water loss, and body temperature in sooty tern. Auk 94:72-79

Magnan A (1911) Morphologie des caecums chez les oiseaux en fonction du régime alimentaire. Ann Sci Nat Zool 14:275-305

Marder J (1973) Temperature regulation in the bedouin fowl (*Gallus domesticus*). Physiol Zool 46:208-217

Marshall EK, Smith HW (1930) The glomerular development of the vertebrate kidney in relation to habitat. Biol Bull 59:135-153

Maumus J (1902) Les caecums des oiseaux. Ann Sci Nat Zool 15:1-148

Mayrs EB (1923) Secretion as a factor in elimination by the bird's kidney. J. Physiol (London) 58:276-287

McBee RH, West GC (1969) Cecal fermentation in the willow ptarmigan. Condor 71:54-58

McFarland LZ (1959) Captive marine birds possessing a functional lateral nasal gland (salt gland). Nature (London) 184:2030-2031

McFarland LZ (1964) Minimal salt load required to induce secretion from the nasal salt glands of sea gulls. Nature (London) 204:1202-1203

McFarland LZ (1965) Influence of external stimuli on the secretory rate of the avian nasal salt gland. Nature (London) 205:391-392

McFarland O, Wright P (1969) Water conservation by inhibition of food intake. Physiol Behav 4:95-99

McGreal RD, Farner DS (1956) Premigratory fat deposition in the Gambel white-crowned sparrow: Some morphologic and chemical observations. Northwest Sci 30:12-23

McNabb FMA (1969a) A comparative study of water balance in three species of quail-I. Water turnover in the absence of temperature stress. Comp Biochem Physiol 28:1045-1058

McNabb FMA (1969b) A comparative study of water balance in three species of quail-II. Utilization of saline drinking solutions. Comp Biochem Physiol 28:1059-1074

McNabb FMA, McNabb RA (1975) Proportions of ammonia, urea, urate and total nitrogen in avian urine and quantitative methods for their analysis on a single urine sample. Poult Sci 54:1498-1505

McNabb FMA, Poulson TL (1970) Uric acid excretion in pigeons, *Columba livia*. Comp Biochem Physiol 33:933-939

McNabb FMA, McNabb RA, Ward JM (1972) The effects of dietary protein content on water requirements and ammonia excretion in pigeons, *Columba livia*. Comp Biochem Physiol 43A:181-185

McNabb FMA, McNabb RA, Steeves HR (1973) Renal mucoid materials in pigeons fed high and low protein diets. Auk 90:14-18

McNabb FMA, McNabb RA, Prather ID, Conner RN, Adkisson CS (1980) Nitrogen excretion by turkey vultures. Condor 82:219–223

McNabb RA (1974) Urate and cation interactions in the liquid and precipitated fractions of avian urine, and speculations on their physico-chemical state. Comp Biochem Physiol 48A:45-54

McNabb RA, McNabb FMA (1975) Urate excretion by the avian kidney. Comp Biochem Physiol 51A:253-258

McNabb RA, McNabb FMA (1977) Avian urinary precipitates: their physical analysis, and their differential inclusion of cations (Ca, Mg) and anions (Cl). Comp Biochem Physiol 56A:621-625

McNabb RA, McNabb FMA, Hinton AP (1973) The excretion of urate and cationic electrolytes by the kidney of the male domestic fowl (*Gallus domesticus*). J Comp Physiol 82:47-57

McWard GW, Scott HM (1961) Sodium requirement of the young chick fed purified diets. Poult Sci 40:1026-1029

Medway W (1958) Total body water in growing domestic fowl by antipyrial dilution technic. Proc Soc Exp Biol Med 99:733-736

Medway W, Kare MR (1957) Water metabolism of the domestic fowl from hatching to maturity. Am J Physiol 190:139-141

Medway W, Kare MR (1959) Water metabolism of the growing domestic fowl with special reference to water balance. Poult Sci 38:631-637

Meier AH (1973) Daily hormone rhythms in the white-throated sparrow. Am Sci 61:184–187

Meier AH, Ferrell BR (1978) Avian endocrinology. In: Brush AH (ed) Chemical zoology, vol 10. Aves. Academic Press, London New York, pp 213-271

Miller MR (1976) Cecal fermentation in mallards in relation to diet. Condor 78:107-111

Milroy TH (1904) The formation of uric acid in birds. J Physiol (London) 30:47-60

Mitchell PC (1901) On the intestinal tract of birds; with remarks on the valuation and nomenclature of zoological characters. Trans Linn Soc Zool (London) 8:173-275

Moldenhauer RR (1970) The effects of temperature on the metabolic rate and evaporative water loss of the sage sparrow *Amphispiza belli nevadensis*. Comp Biochem Physiol 36:579-587

Moldenhauer RR, Wiens JA (1970) The water economy of the sage sparrow, *Amphispiza belli nevadensis*. Condor 72:265-275

Mongin P (1976a) Ionic constituents and osmolality of the small intestinal fluids of the laying hen. Br Poult Sci 17:383-392

Mongin P (1976b) Radioactive polyethyleneglycol: an intestinal water marker in birds. Ann Biol Anim Biochem Biophys 16:631-634

Mongin P, Laage X de (1977) Etudes des mouvements d'eau et d'électrolytes à travers la muqueuse duodenale de la poule pondeuse par perfusion in vivo. C R Acad Sci Paris Ser D 285:225-228

Mongin P, Lacassagne L (1967) Excrétion urinaire chez la poule au moment de la ponte de son premier euf. C R Acad Sci Paris Ser D 264:2479-2480

Mongin P, Sauveur B (1974) Hourly water consumption and egg formation in the domestic fowl. Br Poult Sci 15:361-368

Mongin P, Larbier M, Carbo Baptista N, Licois D, Coudert P (1976) A comparison of the osmotic pressures along the digestive tract of the domestic fowl and the rabbit. Br Poult Sci 17:379-382

Moreau RE, Dolp RM (1970) Fat, water, weights and wing-lenghts of autumn migrants in transit on the northwest coast of Egypt. Ibis 112:209-228

Mortensen A, Tindall A (1978) Uric acid metabolism in the caeca of grouse (*Lagopus lagopus.*) J Physiol (London) 284:159P-160P

Morton ML (1979) Fecal sac ingestion in the mountain white-crowned sparrow. Condor 81:72-77

Mouchette R, Cuypers Y (1959) Etude de la vascularisation du rein de coq. Arch Biol 69:577-590

Müller S (1922) Zur Morphologie des Oberflächenreliefs der Rumpfdarmschleimhaut bei den Vögeln. Jena Z Naturwiss 58:533-606

Mulkey GJ, Huston TM (1967) The tolerance of different ages of domestic fowl to body water loss. Poult Sci 46:1564-1569

Munsick RA (1964) Neurohypophysial hormones of chickens and turkeys. Endocrinol 75:104-112

Munsick RA, Sawyer WH, Van Dyke HB (1960) Avian neurohypophysial hormones: pharmacological properties and tentative identification. Endocrinology 66:860-871

Murrish DE, Schmidt-Nielsen K (1970) Water transport in the cloaca of lizards: Active or passive? Science 170:324-326

Myrcha A, Pinowski J (1969) Variations in the body composition and caloric value of nestling tree sparrows (*Passer m. montanus L.*). Bull Acad Pol Sci 17:475-480

Myrcha A, Pinowski J, Tomek T (1973) Variations in the water and ash contents and in the caloric value of nestling starlings (*Sturnus vulgaris L.*) during their development. Bull Acad Pol Sci 21:649-655

Nagra CL, Birnie JG, Baum GJ, Meyer RK (1963) The role of the pituitary in regulating steroid secretion by the avian adrenal. Gen Comp Endocrinol 3:274-280

Nakajima T, Khosla MC, Sakakibara S (1978) Comparative biochemistry of renins and angiotensins in the vertebrates. Jpn Heart J 19:799-805

Nakayama T, Nakajima T, Sokabe H (1973) Comparative studies on angiotensins. III. Structure of fowl angiotensin and its identification by DNS-method. Chem Pharm Bull 21:2085-2087

Neal CM, Lustick SI (1973) Energetics and evaporative water loss in the short-tailed shrew, *Blarina brevicauda*. Physiol Zool 46:180-185

Nechay BR, Nechay L (1959) Effects of probenecid sodium, salicylate, 2.4-dinitrophenol and pyrazinamide on renal secretion of uric acid in chickens. J Pharmacol Exp Ther 26:291-295

Nechay BR, Boyarsky S, Catacutan-Labay P (1969) Rapid migration of urine into intestine chickens. Comp Biochem Physiol 26:369-370

Nesbeth WG, Douglas CR, Harms RH (1976) Response of laying hens to a low salt diet. Poult Sci 55:2128-2133

Newell GW, Shaffner CS (1950) Blood volume determinations in chickens. Poult Sci 29:78-87

Niezgoda G, Rzasa J (1971) Blood levels of vasotocin in birds. Bull Acad Pol Sci Biol Cl II 19:359-361

Niezgoda J (1975) Changes in blood arginine vasotocin level after osmotic stimuli in domestic fowl. Acta Physiol Pol 26:591-597

Nirmalan GP, Robinson GA (1972) Effect of age, sex, and egg laying on the total erythrocyte volume ($^{51}CrO_4$ label) and the plasma volume (^{125}I-serum albumin label) of japanese quail. Can J Physiol Pharmacol 50:6-10

Nisbet ICT (1963) Weight-loss during migration. Part II: review of other estimates. Bird-Banding 34:139-159

Nisbet ICT, Drury WH, Baird J (1963) Weight-loss during migration. Part I: Deposition and consumption of fat by the blackpoll warbler, *Dendroica striata*. Bird-Banding 34:107-138

Nishimura H (1978) Physiological evolution of the renin-angiotensin system. Jpn Heart J 19:806-822

Nott H, Combs GF (1969) Sodium requirement of the chick. Poult Sci 48:660-665

Nys Y, Sauveur B, Lacassagne L, Mongin P (1976) Food, calcium and water intakes by hens lit continuously from hatching. Br Poult Sci 17:351-358

O'Dell BL, Woods WD, Laerdal OA, Jeffay AM, Savage JE (1960) Distribution of the major nitrogenous compounds and amino acids in chicken urine. Poult Sci 39:426-432

Odlind B (1977) Blood flow distribution in the renal portal system of the intact hen. A study of a venous system using microspheres. Acta Physiol Scand 102:342-356

Odum EP, Rogers DT, Hicks DL (1964) Homeostasis of the nonfat components of migrating birds. Science 143:1037-1039

Oelofsen BW (1973) Renal function in the pengiun (*Spheniscus demersus*) with special reference to the role of the renal portal system and renal portal valves. Zool Afr 8:41-62

Oelofsen BW (1977) The renal portal valves of the ostrich *Struthio camelus* S Afr J Sci 73:57-58

Ogawa M, Sokabe H (1971) The macula densa site of avian kidney. Z Zellforsch 120:29-36

Ohmart RD (1972) Physiological and ecological observations concerning the salt-secreting nasal glands of the roadrunner. Comp Biochem Physiol 43A:311-316

Ohmart RD, Lasiewski RC (1971) Roadrunners: energy conservation by hypothermia and absorption of sunlight. Science 172:67-69

Ohmart RD, Smith EL (1970) Use of sodium chloride solutions by the Brewer's sparrow and tree sparrow. Auk 87:329-341

Ohmart RD, Smith EL (1971) Water deprivation and use of sodium chloride solutions by vesper sparrows (*Pooecetes gramineus*). Condor 82:364-366

Ohmart RD, Chapman TE, McFarland LZ (1970) Water turnover in roadrunners under different environmental conditions. Auk 87:787-793

Oksche A, Farner DS (1974) In: Brodal A (ed) Advances in anatomy, embryology and cell biology. Springer, Berlin Heidelberg New York

Oksche A, Farner DS, Serventy DL, Wolff F, Nicholls CA (1963) The hypothalamo-hypophysial neurosecretory system of the zebra finch, *Taeniopygia castanotis*. Z Zellforsch 58:846-914

Oksche A, Laws DF, Kanemoto FI, Farner DS (1959) The hypothalamo-hypophysial neurosecretory system of the white-crowned sparrow, *Zonotrichia leucophrys gambelii*. Z Zellforsch 51:1-42

Olson C, Mann FC (1935) The physiology of the cecum of the domestic fowl. J Am Vet Med Assoc 87:151-159

Orloff J, Davidson DG (1959) The mechanism of potassium excretion in the chicken. J Clin Invest 38:21-30

Osbaldiston GW (1969a) Alteration of glomerular filtration rate in the chicken during intravenous dextrose infusion. Br Vet J 125:645-652

Osbaldiston GW (1969b) Water and electrolyte balance studies of birds showing "wet droppings". Br Vet J 125:653-663

Owen EE, Robinson RR (1964) Urea production and excretion by the chicken kidney. Am J Physiol 206:1321-1325

Pang CY, Campbell LD, Phillips GD (1978) Pathophysiological changes in plasma and body composition of young poults fed a sodium-deficient diet. Can J Anim Sci 58:597-604

Parhon CC, Bârză E (1967) Recherches sur l' absorption caecale chez les Oiseaux domestiques. Rev Roum Biol Ser Zool 12:109-115

Paulson GD (1969) An improved method for separate collection of urine, feces, and expiratory gases from the mature chicken. Poult Sci 48:1331-1336

Peaker M (1971) Intracellular concentrations of sodium, potassium and chloride in the salt-gland of the domestic goose and their relation to the secretory mechanism. J Physiol (London) 213:399-410

Peaker M (ed) (1975) Avian physiology. Symp Zool Soc London, vol 35. Academic Press, London New York, pp 1-377

Peaker M (1978) Do osmoreceptors or blood volume receptors initiate salt-gland secretion in birds? J Physiol (London) 276:66P-67P

Peaker M (1979) Control mechanism in Birds. In: Gilles R (ed) Control Mechanism of osmoregulation in animals. John Wiley and Sons, Chichester, pp 323-348

Peaker M, Linzell JL (1975) Salt glands in birds and reptiles. Cambridge University Press, pp 1–307

Peaker M, Wright A, Peaker SJ, Phillips JG (1968) Absorption of tritiated water by the cloaca of the domestic duck. *Anas platyrhynchos.* Physiol Zool 41:461-465

Peaker M, Peaker SJ, Phillips JG, Wright A (1971) The effects of corticotrophin, glucose and potassium chloride on secretion by the nasal salt gland of the duck, *Anas platyrhynchos.* J Endocrinol 50:293-299

Peaker M, Peaker SJ, Hanwell A, Linzell JL (1973) Sensitivity of the receptors for salt-gland secretion in the domestic duck and goose. Comp Biochem Physiol 44A:41-46

Phillips JG, Ensor DM (1972) The significance of environmental factors in the hormone mediated changes of nasal (salt) gland activity in birds. Gen Comp Endocrinol Suppl 3:393-404

Phillips JG, Holmes WN, Butler DG (1961) The effect of total and subtotal adrenalectomy on the renal and extra-renal response of the domestic duck (*Anas platyrhynchos*) to saline loading. Endocrinology 69:958-969

Pitts RF (1938) The excretion of phenol red by the chicken. J Cell Comp Physiol 11:99-115

Pitts RF, Korr IM (1938) The excretion of urea by the chicken. J Cell Comp Physiol 11:117-122

Polin D, Wynosky ER, Loukides M, Porter CC (1967) A possible urinary backflow to ceca revealed by studies on chicks with artificial anus and fed Amprolium-C[14] or Thiamine-C[14]. Poult Sci 46:88-93

Porter P (1963a) Physico-chemical factors involved in urate calculus formation. I. Solubility. Res Vet Sci 4:580-591

Porter P (1963b) Physico-chemical factors involved in urate calculus formation. II. Colloidal flocculation. Res Vet Sci 4:592-602

Portman OW, McConnell KP, Rigdon RH (1952) Blood volume of ducks using human serum albumin labeled with radioiodine. Proc Soc Exp Biol Med 81:599-601

Poulson TL (1965) Countercurrent multipliers in avian kidneys. Science 148:389-391

Poulson TL (1969) Salt and water balance in seaside and sharp-tailed sparrows. Auk 86:473-489

Poulson TL, Bartholomew GA (1962a) Salt balance in the savannah sparrow. Physiol Zool 35:109-119

Poulson TL, Bartholomew GA (1962b) Salt utilization in the house finch. Condor 64:245-252

Poulson TL, McNabb FMA (1970) Uric acid: the main nitrogenous excretory product of birds. Science 170:98

Purdue JR, Haines H (1977) Salt water tolerance and water turnover in the snowy plover. Auk 94:248-255

Radeff T (1928) Über die Rohfaserverdauung beim Huhn und die hierbei dem Blinddarm zukommende Bedeutung. Biochem Z 193:192-196

Ralph CL (1960) Polydipsia in the hen following lesions in the supraoptic hypothalamus. Am J Physiol 198:528-530

Rehberg PB (1926) Studies on kidney function. Thesis, Copenhagen

Rennick BR, Gandia H (1954) Pharmacology of smooth muscle valve in renal portal circulation of birds. Proc Soc Exp Biol 85:234-236

Rennick BR, Kandel A, Peters L (1956) Inhibition of renal tubular excretion of tetraethylammonium and N-methylnicotinamide by basic cyanide dyes. J Pharmacol Exp Ther 118:204-219

Resko JA, Norton HW, Nalbandov AV (1964) Endocrine control of the adrenal in chickens. Endocrinology 75:192-200

Richardson CE, Watts AB, Wilkinson WS, Dixon JM (1960) Techniques used in metabolism studies with surgically modified hens. Poult Sci 432-440

Ricklefs RE (1968) Weight recession in nestling birds. Auk 85:30-35

Rijke AM (1972) The water-holding mechanism of sandgrouse feathers. J Exp Biol 56:195-200

Röseler M (1929) Die Bedeutung der Blinddärme des Haushuhnes für die Resorption der Nahrung und Verdauung der Rohfaser. Z Tierzucht Zuechtungsbiol 13:281-310

Romanoff AL, Romanoff AJ (1967) Biochemistry of the avian ambryo. A quantitative analysis of prenatal development. John Wiley and Sons, New York, pp 1-398

Ross E (1979) The effect of waters and sodium on the chick requirement for dietary sodium. Poult Sci 58:626-630

Rothchild I (1947) The artificial anus in the bird. Poult Sci 26:157-162

Rothwell B (1974) Perfusion fixation of the kidney of the domestic fowl. J Microsc 100:99-104

Rounsevell D (1970) Salt excretion in the Australian pipit, *Anthus novaeseelandiae* (Aves: motacillidae). Aust J Zool 18:373-377

Ruch FE, Hughes MR (1975) The effects of hypertonic sodium chloride injection on body water distribution in ducks (*Anas platyrhynchos*), gulls (*Larus glaucescens*), and roosters (*Gallus domesticus*). Comp Biochem Physiol 52A:21-28

Rzasa J (1972) Effect of vasotocin and oxytocin on contractility of the hen oviduct in vitro. Acta Physiol Pol 23:735-745

Rzasa J (1978) Effects of arginine vasotocin and prostaglandin E1 on the hen uterus. Prostaglandins 16:357-373

Rzasa J, Ewy Z (1971) Effect of vasotocin and oxytocin on intrauterine pressure in the hen. J Reprod Fertil 25:115-116

Rzasa J, Niezgoda J (1969) Effects of the neurohypophysial hormones on sodium and potassium level in the hen's blood. Bull Acad Pol Sci 17:585-588

Rzasa J, Niezgoda J, Kahl S (1974) The effect of arginine vasotocin on blood volume, plasma protein and electrolyte concentrations in the cockerel, *Gallus domesticus*. Br Poult Sci 15:261-265

Salem MHM, Norton HW, Nalbandov AV (1970a) A study of ACTH and CRF in chickens. Gen Comp Endocrinol 14:270-280

Salem MHM, Norton HW, Nalbandov AV (1970b) The role of vasotocin and of CRF in ACTH release in the chicken. Gen Comp Endocrinol 14:281-289

Salt GW (1964) Respiratory evaporation in birds. Biol Rev 39:113-136

Sandor T, Fazekas AG, Robinson BH (1976) The biosynthesis of corticosteroids throughout the vertebrates. In: Chester Jones I, Henderson IW (eds) General, comparative and clinical endocrinology of the adrenal cortex. Academic Press, London New York, pp 25-142

Sandor T, Mehdi AZ, Fazekas AG (1977) Corticosteroid-binding macromolecules in the saltactivated nasal gland of the domestic duck. Gen Comp Endocrinol 32:348-359

Sanner E (1965) Studies on biogenic amines and reserpine induced block of the diuretic action of hydrochlorothiazide and theophylline in the chicken. Acta Pharmacol Toxical 22:Suppl 1

Sapirstein LA, Hartman FA (1959) Cardiac output and its distribution in the chicken. Am J Physiol 196:751-752

Sauveur B (1969) Acidoses metaboliques experimentales chez la poule pondeuse. Ann Biol Anim Biochem Biophys 9:379-391

Sauveur B (1973) Alcalose induite par le nitrate chez la poule et caracteristiques de l'oeuf. Ann Biol Anim Biochem Biophys 13:611-626

Sawyer WH (1967) Evolution of antidiuretic hormones and their functions. Am J Med 42:678-686

Schade H, Boden E (1913) Über die Anomalie der Harnsäurelöslichkeit (kolloide Harnsäure). Z Physiol Chem 83:347-380

Scheiber AR, Dziuk HE (1969) Water ingestion and excretion in turkeys with rectal fistula. J Appl Physiol 26:277-281

Scheiber AR, Dziuk HE, Duke GE (1969) Effects of a chronic colostomy in turkeys. Poult Sci 48:2179-2182

Scheid P (1979) Mechanisms of gas exchange in bird lungs. Rev Physiol Biochem Pharmacol 86:137-186

Schildmacher H (1932) Über den Einfluss des Salzwassers auf die Entwicklung der Nasendrüsen. J Ornithol 80:293-299

Schlotthauer CF, Essex HE, Mann FC (1933) Cecal occlusion in the prevention of blackhead (enterohepatitis) in turkeys. J Am Vet Med Assoc 83:218-228

Schmidt-Nielsen B (1964) Organ systems in adaptation: The excretory system. In: Dill DB, Adolph EF, Wilder CG (eds) Handb Physiol Sect 4. Am Physiol Soc, Washington 1964

Schmidt-Nielsen B (1976) Intracellular concentrations of the salt gland of the herring gull, *Larus argentatus*. Am J Physiol 230:514-521

Schmidt-Nielsen B, O'Dell R (1961) Structure and concentrating mechanism in the mammalian kidney. Am J Physiol 200:1119-1124

Schmidt-Nielsen B, Skadhauge E (1967) Function of the excretory system of the crocodile (*Crocodylus acutus*). Am J Physiol 212:973-980

Schmidt-Nielsen K (1960) The salt-secreting gland of marine birds. Circulation 21:955-967

Schmidt-Nielsen K (1978) Introduction. In: Barker Jørgensen C, Skadhauge E (eds) Osmotic and volume regulation. Munksgaard, Copenhagen, pp 299-309

Schmidt-Nielsen K, Fänge R (1958) The function of the salt gland in the brown pelican. Auk 75:282-289

Schmidt-Nielsen K, Kim YT (1964) The effect of salt intake on the size and function of the salt gland of ducks. Auk 81:160-172

Schmidt-Nielsen K, Schmidt-Nielsen B, Jarnum SA, Houpt TR (1957) Body temperature of the camel and its relation to water economy. Am J Physiol 188:103-112

Schmidt-Nielsen K, Jørgensen CB, Osaki H (1958) Extrarenal salt excretion in birds. Am J Physiol 193:101-107

Schmidt-Nielsen K, Kanwisher J, Lasiewski RC, Cohn JE, Bretz WL (1969) Temperature regulation and respiration in the ostrich. Condor 71:341-352

Schmidt-Nielsen K, Borut A, Lee P, Crawford E (1963) Nasal salt excretion and the possible function of the cloaca in water conservation. Science 142:1300-1301

Schmidt-Nielsen K, Hainsworth FR, Murrish DE (1970) Counter-current heat exchange in the respiration passages: effect on water and heat balance. Respir Physiol 9:263-276

Schütte KH (1973) The composition of ostrich urine. S Afr J Sci 69:56-57

Schultz SG, Curran PF, Chez RA, Fuisz RE (1967) Alanine and sodium fluxes across mucosal border of rabbit ileum. J Gen Physiol 50:1241-1260

Schwarz D (1962) Untersuchungen zur biologischen Bedeutung der Salzdrüsen bei freilebenden Sturmmöwen (Larus canus L.) J Ornithol 103:180-186

Schwarz D (1965) Das Verhalten von Lariden gegenüber Salzlösungen von verschiedener Konzentration. Beitr z Vogelkde 11:359

Scothorne RJ (1959) On the response of the duck and the pigeon to intravenous hypertonic saline solutions. Q J Expl Physiol 44:200-207

Scott ML, Tienhoven A van, Holm ER, Reynolds RE (1960) Studies on the sodium chlorine and iodine requirements of young pheasants and quail. J Nutr 71:282-288

Seley H (1943) Production of nephrosclerosis in the fowl by NaCl. J Am Vet Med Assoc 103:140-143

Sell JL, Balloun SL (1961) Nitrogen retention and nitrogenous urine components of growing cockerels as influenced by diethylstilbestrol, methyl testosterone and porcine growth hormone. Poult Sci 40:1117-1129

Serventy DL (1972) Biology of desert birds. In: Farner DS, King JR (eds) Avian biology, vol I, Chap 7. Academic Press, London New York, pp 287-339

Serventy DL, Whittell HM (1967) Birds of Western Australia. Lamb Publications Perth, WA, pp 1-440

Shannon JA (1938a) The excretion of exogenous creatinine by the chicken. J Cell Comp Physiol 11:123-134

Shannon JA (1938b) The excretion of uric acid by the chicken. J Cell Comp Physiol 11:135-148

Sharp PJ, Beuving G (1978) Role of corticosterone in ovulatory cycle of hen. J Endocrinol 78:195-200

Sharpe NC (1912) On the secretion of urine in birds. Am J Physiol 31:75-84

Sharpe NC (1923) On absorption from the cloaca in birds. Am J Physiol 66:209-213

Shirley HV, Nalbandov AV (1956) Effects of neurohypophysectomy in domestic chickens. Endocrinology 58:477-483

Shoemaker VH (1972) Osmoregulation and excretion in birds. In: Farner DS, King JR (eds) Avian biology, vol 2, Chap 9. Academic Press, London New York, pp 527-574

Siller WG (1971) Structure of the kidney. In: Bell DJ, Freeman BM (eds) Physiology and biochemistry of the domestic fowl, vol I, Chap 8. Academic Press, London New York, pp 197-231

Siller WG, Hindle RM (1969) The arterial blood supply to the kidney of the fowl. J Anat 104:117-135

Silva P, Stoff J, Field M, Fine L, Forrest JN, Epstein FH (1977) Mechanism of active chloride secretion by shark rectal gland: role of Na-K-ATPase in chloride transport. Am J Physiol 233:F298-F306

Simensen E, Olson LD, Vanjonack WJ, Johnson HD, Ryan MP (1978) Determination of corticosterone concentration in plasma of turkeys using radioimmunoassay. Poult Sci 57:1701-1704

Simon-Oppermann C, Simon E, Jessen C, Hammel HT (1978) Hypothalamic thermosensitivity in conscious Pekin ducks. Am J Physiol 235:R130-R140

Simon-Oppermann C, Hammel HT, Simon E (1979) Hypothalamic temperature and osmoregulation in the Pekin duck. Pflügers Arch Ges Physiol 378:213-221

192

Skadhauge E (1964) Effects of unilateral infusion of arginine-vasotocin into the portal circulation of the avian kidney. Acta Endocrin 47:321-330

Skadhauge E (1967) In vivo perfusion studies of the water and electrolyte resorption in the cloaca of the fowl (*Gallus domesticus*). Comp Biochem Physiol 23:483-501

Skadhauge E (1968) Cloacal storage of urine in the rooster. Comp Biochem Physiol 24:7-18

Skadhauge E (1969) Activités biologiques des hormones neuro-hypophysaires chez les Oiseaux et les Reptiles. Colloq Int C N R S 177:63-68

Skadhauge E (1973) Renal and cloacal salt and water transport in the fowl (*Gallus domesticus*). Dan Med Bull 20: Suppl. 1,1-82

Skadhauge E (1974a) Renal concentrating ability in selected West Australian birds. J Exp Physiol 61:269-276

Skadhauge E (1974b) Cloacal resorption of salt and water in the galah (*Cacatua roseicapilla*). J Physiol (London) 240:763-773

Skadhauge E (1975) Renal and cloacal transport of salt and water. Symp Zool Soc London 35:97-106

Skadhauge E (1976a) Cloacal absorption of urine in birds. Comp Biochem Physiol 55A:93-98

Skadhauge E (1976b) Water conservation in xerophilic birds. Isr J Med Sci 12:732-739

Skadhauge E (1977) Solute composition of the osmotic space of ureteral urine in dehydrated chickens (*Gallus domesticus*). Comp Biochem Physiol 56A:271-274

Skadhauge E (1978) Hormonal regulation of salt and water balance in granivorous birds. In: Assenmacher I, Farner DS (eds) Environmental Endocrinology, Springer, Berlin Heidelberg New York, pp 222-229

Skadhauge E (1979) Aldosterone effects on lower intestinal sodium transport. Colloq INSERM 85:121-128

Skadhauge E, Bradshaw SD (1974) Saline drinking and cloacal excretion of salt and water in the zebra finch. Am J Physiol 227:1263-1267

Skadhauge E, Dawson TJ, (1980a) Excretion of several ions and water in a xerophilic parrot. Comp Biochem Physiol 65A:325-330

Skadhauge E, Dawson TJ (1980b) In vitro studies of sodium transport across the lower intestine of a desert parrot. Am J Physiol 239:R285-290

Skadhauge E, Kristensen K (1972) An analogue computer simulation of cloacal resorption of salt and water from ureteral urine in birds. J Theor Biol 35:473-487

Skadhauge E, Schmidt-Nielsen B (1967a) Renal function in domestic fowl. Am J Physiol 212:793-798

Skadhauge E, Schmidt-Nielsen B (1967b) Renal medullary electrolyte and urea gradient in chickens and turkeys. Am J Physiol 212:1313-1318

Skadhauge E, Thomas DH (1979) Transepithelial transport of K, NH_4, inorganic phosphate and water by hen (*Gallus domesticus*) lower intestine (colon and coprodeum) perfused luminally in vivo. Pflügers Arch 379:327-343

Skadhauge E, Herd RM, Dawson TJ (1980) Renal excretion, and cloacal absorption of several ions and water in the emu (*Dromaius novae-hollandiae*, Aves). Proc Int Union Physiol Sci 14:707

Skar H-J, Hagen A, Oestbye E (1972) Caloric values, weight, ash and water content of the body of the meadow pipit (*Anthus pratensis* (L.)). Norw J Zool 20:51-59

Smith DP (1972) An investigation of interrelated factors affecting the flow and concentration of salt-induced nasal secretion in the duck, *Anas platyrhynchos*. Comp Biochem Physiol 43A:1003-1017

Smith HW (1953) From fish to philosoper. Little, Brown and Co, Boston

Smyth M, Bartholomew GA (1966a) Effects of water deprivation and sodium chloride on the blood and urine of the mourning dove. Auk 83:597-602

Smyth M, Bartholomew GA (1966b) The water economy of the black-throated sparrow and the rock wren. Condor 68:447-458

Sokabe H (1974) Phylogeny of the renal effects of angiotensin. Kidney Int 6:263-271

Sokabe H, Ogawa M (1974) Comparative studies of the juxtaglomerular apparatus. Int Rev Cytol 37:271-327

Sossinka R (1970) Domestikationserscheinungen beim Zebrafinken *Taeniopygia guttata castanotis* (Gould). Zool Jahrb Syst 97:455-521

Sossinka R (1972) Langfristiges Durstvermögen wilder und domestizierter Zebrafinken (*Taeniopygia guttata castanotis*) (Gould). J Ornithol 113:418-426

Spanner R (1925) Der Pfortaderkreislauf in der Vogelniere. Morphol J 54:560-632

Spanner R (1939) Die Drosselklappe der veno-venösen Anastomose und ihre Bedeutung für den Abkürzungskreislauf im portocavalen System des Vogels; zugleich ein Beitrag zur Kenntnis epitheloiden Zellen. Z Anat Entw Gesch 109:443-492

Sperber I (1948) The excretion of some glucuronic acid derivatives and phenol sulphuric esters in the chicken. K. Lantbohoegsk Annlr 15:317-349

Sperber I (1949) Investigations on the circulatory system of the avian kidney. Zool Bidrag 27:429-448

Sperber I (1960) Excretion. In: Marshall AJ (ed) Biology and physiology of birds, vol I. Academic Press, London New York, pp 469-492

Staaland H (1967) Anatomical and physiological adaptations of the nasal glands in charadriiformes birds. Comp Biochem Physiol 23:933-944

Stainer IM, Holmes WN (1969) Some evidence for the presence of a corticotrophin releasing factor (CRF) in the duck (*Anas platyrhynchos*). Gen Comp Endocrinol 12:350-359

Stewart DJ (1972) Secretion by salt gland during water deprivation in the duck. Am J Physiol 223:384-386

Stewart DJ, Holmes WN, Fletcher G (1969) The renal excretion of nitrogenous compounds by the duck (*Anas platyrhynchos*) maintained on freshwater and on hypertonic saline. J Exp Biol 50:527-539

Sturkie PD (ed) (1976) Avian physiology, Springer advanced texts in life sciences. Springer, Berlin Heidelberg New York, pp 1-400

Sturkie PD, Joiner WP (1959a) Effects of cloacal cannulation on feed and water consumption in chickens. Poult Sci 38:30-32

Sturkie PD, Joiner WP (1959b) Effects of foreign bodies in cloaca and rectum of the chicken on feed consumption. Am J Physiol 197:1337-1338

Sturkie PD, Lin Y-C (1966) Release of vasotocin and oviposition in the hen. J Endocrinol 35:325-326

Sturkie PD, Lin Y-C (1967) Further studies on oviposition and vasotocin release in the hen. Poult Sci 46:1591-1592

Sturkie PD, Dirner G, Gister R (1977) Shunting of blood from the renal portal to the hepatic portal circulation of chickens. Comp Biochem Physiol 58A:213-215

Sturkie PD, Dirner G, Gister R (1978) Role of renal portal valve in the shunting of blood flow in renal and hepatic circulations of chickens. Comp Biochem Physiol 59C:95-96

Sunde ML, Cravens WW, Elvehjem CA, Halpin JG (1950) The effect of diet and caecectomy on the intestinal synthesis of biotin in the mature fowl. Poult Sci 29:10-14

Svendsen C, Skadhauge E (1976) Renal functions in hens fed graded dietary levels of ochratoxin A. Acta Pharmacol Toxicol 38:186-194

Sykes AH (1960) The renal clearance of uric acid and p-aminohippurate in the fowl. Res Vet Sci 1:308-314

Sykes AH (1962) The excretion of urea following its infusion through the renal portal system of the fowl. Res Vet Sci 3:183-185

Sykes AH (1971) Formation and composition of urine. In: Bell DJ, Freeman BM (eds) Physiology and biochemistry of the domestic fowl, vol I, Chap 9. Academic Press, London New York, pp 233-278

Szalágyi K, Kriwuscha A (1914) Untersuchungen über die chemische Zusammensetzung und die physikalischen Eigenschaften des Enten- und Hühnerharnes. Biochem Z 66:122-138

Szwykowska MM (1969) Seasonal changes of the caloric value and chemical composition of the body of the partridge (*Perdix perdix L.*). Ekol Pol 17:795-809

Takei Y (1977) Angiotensin and water intake in the Japanese quail (*Coturnix coturnix japonica*). Gen Comp Endocrinol 31:364-372

Tanaka K, Nakajo S (1962) Participation of neurohypophysial hormone in oviposition in the hen. Endocrinology 70:453-458

Tasaki I, Okumura J (1964) Effect of protein level of diet on nitrogen excretion in fowls. J Nutr 83:34-38

Taylor AA, Davis JO, Breitenbach RP, Hartroft PM (1970) Adrenal steroid secretion and a renal-pressor system in the chicken (*Gallus domesticus*). Gen Comp Endocrinol 14:321-333

Taylor CR (1972) The desert gazelle: A paradox resolved. Symp Zool Soc London 31:215-227

Taylor TG, Kirkley J (1967) The absorption and excretion of minerals by laying hens in relation to egg shell formation. Br Poult Sci 8:289-295

Tchang LK (1923) Recherches histologiques sur la structure du rein des oiseaux. Thesis, Lyon

194

Technau G (1936) Die Nasendrüse der Vögel. Zugleich ein Beitrag zur Morphologie der Nasen-Höhle. J Ornithol 84:511-617

Templeton JR (1964) Nasal salt excretion in terrestrial lizard. Comp Biochem Physiol 11:223-229

Thomas DH (1973) Control of active ion and water transfer in the chicken rectum: in vitro effects of aldosterone. J Endocrinol 59:33

Thomas DH, Phillips JG (1975a) Studies in avian adrenal steroid function. I: Survival and mineral balance following adrenalectomy in domestic ducks (*Anas platyrhynchos L.*). Gen Comp Endocrinol 26:394-403

Thomas DH, Phillips JG (1975b) Studies in avian adrenal steroid function. II: Chronic adrenalectomy and the turnover of $[^3H]_2O$ in domestic ducks (*Anas platyrhynchos L.*). Gen Comp Endocrinol 26:404-411

Thomas DH, Phillips JG (1975c) Studies in avian adrenal steroid function. III: Adrenalectomy and the renal-cloacal response in water-loaded domestic ducks (*Anas platyrhynchos L.*). Gen Comp Endocrinol 26:412–419

Thomas DH, Phillips JG (1975d) Studies in avian adrenal steroid function. IV: Adrenalectomy and response of domestic ducks (*Anas platyrhynchos L.*) to hypertonic NaCl loading. Gen Comp Endocrinol 26:427-439

Thomas DH, Phillips JG (1975e) Studies in avian adrenal steroid function. V: Hormone kinetics and the differentiation of mineralcorticoid and glucocorticoid effects. Gen Comp Endocrinol 26:440-450

Thomas DH, Robin AP (1977) Comparative studies of thermoregulatory and osmoregulatory behaviour and physiology of five species of sandgrouse (Aves: Pterocliidae) in Morocco. J Zool 183:229-249

Thomas DH, Skadhauge E (1979a) Dietary Na^+ effects on transepithelial transport of NaCl by hen (*Gallus domesticus*) lower intestine (colon and coprodeum) perfused luminally in vivo. Pflügers Arch 379:229–236

Thomas DH, Skadhauge E (1979b) Chronic aldosterone therapy and the control of transepithelial transport of ions and water by the colon and coprodeum of the domestic fowl (*Gallus domesticus*) in vivo. J Endocrinol 83:239-250

Thomas DH, Skadhauge E, Read MW (1975) Steroid effects on gut function in birds. Biochem Soc Trans 3:1164-1168

Thomas DH, Skadhauge E, Read MW (1979) Acute effects of aldosterone on water and electrolyte transport by the colon and coprodeum of the domestic fowl (*Gallus domesticus*) in vivo. J Endocrinol 83:229-237

Thomas DH, Jallagéas M, Munck BG, Skadhauge E (1980) Aldosterone effects on electrolyte transport of the lower intestine (coprodeum and colon) of the fowl (*Gallus domesticus*) in vitro. Gen Comp Endocrinol 40:44-51

Thornburn CC, Willcox JS (1965) The caeca of the domestic fowl and digestion of the crude fibre complex. Br Poult Sci 6:23-31

Timet D, Mitin V, Herak M, Emanovic D, Kraljevic P (1971) L'absorption des acides gras volatils dans les caeca de la poule. J Physiol (Paris) 63:296 A-297 A

Torre-Bueno JR (1978) Evaporative cooling and water balance during flight in birds. J Exp Biol 75:231-236

Tucker VA (1968) Respiratory exchange and evaporative water loss in the flying budgerigar. J Exp Biol 48:67-87

Tyler C (1946) Studies in the absorption and excretion of certain minerals by poultry. J Agric Sci 36:275-282

Uemura H (1964) Effects of water deprivation on the hypothalamo-hypophysial neurosecretory system of the grass parakeet, *Melopsittacus undulatus*. Gen Comp Endocrinol 4:193-198

Ussing HH (1949) The distinction by means of tracers between active transport and diffusion. The transfer of iodide across the isolated frog skin. Acta Physiol Scand 19:43-56

Ussing HH, Zerahn K (1951) Active transport of sodium as the source of electric current in the short-circuited isolated frog skin. Acta Physiol Scan 23:110-127

Van Tyne J, Berger AJ (1959) Fundamentals of ornithology. John Wiley and Sons, New York pp 1–624

Vinson GP, Whitehouse BJ, Goddard C, Sibley CP (1979) Comparative and evolutionary aspects of aldosterone secretion and zona glomerulosa function. J Endocrinol 81:5P-24P

Vogel G, Stoeckert I, Kroger W, Dobberstein I (1965) Harn und Harnbereitung bei terrestrisch lebenden Vögeln. – Untersuchungen am Truthuhn (*Meleagris pavo L.*). Zentralbl Veterinärmed 12:132–160

Vogt H, Nezel K, Rauch W, Stute K (1970) Bedarf und Verträglichkeit von Natrium bei Mastküken und Legehennen. Arch Geflügelkde 34:116-122

Volle RL, Green RE, Peters L, Handschumacher RE, Welch AD (1962) Renal tubular excretion studies with pyramidine derivatives and analogs. J Pharmacol Exp Ther 136:353-360

Wada M, Kobayashi H, Farner DS (1975) Induction of drinking in the white-crowned sparrow, *Zonotrichia leucophrys gambelii*, by intracranial injection of angiotensin II. Gen Comp Endocrinol 26:192-197

Walsberg GE, Campbell GS, King JR (1978) Animal coat color and radiative heat gain: A re-evaluation. J Comp Physiol 126:211-222

Walter A, Hughes MR (1978) Total body water volume and turnover rate in fresh water and sea water adapted glaucous-winged gulls, *Larus glaucescens*. Comp Biochem Physiol 61:233-237

Ward JM, McNabb RA, McNabb FMA (1975a) The effects of changes in dietary protein and water availability on urinary nitrogen compounds in the rooster, *Gallus domesticus*-I. Urine flow and the excretion of uric acid and ammonia. Comp Biochem Physiol 51A: 165-169

Ward JM, McNabb RM, McNabb FMA (1975b) The effect of changes in dietary protein and water availability on urinary nitrogen compounds in the rooster, *Gallus domesticus*-II. Diurnal patterns in urine flow rates, and urinary uric acid and ammonia concentrations. Comp Biochem Physiol 51A:171-174

Weathers WW, Caccamise DF (1975) Temperature regulation and water requirements of the monk parakeet, *Myiopsitta monachus*. Oecologia (Berlin) 18:329-342

Weathers WW, Schoenbaechler DC (1976) Regulation of body temperature in the budgerigar, *Melopsittacus undulatus*. Aust J Zool 24:39-47

Weathers WW, Shapiro CJ, Astheimer LB (1980) Metabolic responses of Cassin's finches (*Carpodacus cassinii*) to temperature. Comp Biochem Physiol 65A:235-238

Weiss HS (1958) Application to the fowl of the antipyrine dilution technique for the estimation of body composition. Poult Sci 37:484-489

Wels A, Schnappauf HP, Horn V (1967) Blutvolumenbestimmung bei Hühnern mit Cr^{51} und T-1824. Zentralbl Veterinärmed Reihe A 14:741–746

Weyrauch HM, Roland SI (1958) Electrolyte absorption from fowls cloaca: resistance to hyperchloremic acidosis. J Urol 79:255-263

Whitehead CC, Shannon DWF (1974) The control of egg production using a low-sodium diet. Br Poult Sci 15:429-434

Wiener H (1902) Über synthetische Bildung der Harnsäure im Tierkörper. Beitr Chem Physiol Pathol 2:42-85

Williams CK, Main AR (1976) Ecology of Australian chats (*Epthianura*), Gould): seasonal movements, metabolism and evaporative water loss. Aust J Zool 24:397-416

Williams CK, Main AR (1977) Ecology of Australian chats (*Epthianura* Gould): Aridity, electrolytes and water economy. Aust J Zool 25:673-691

Willoughby EJ (1966) Water requirements of the ground dove. Condor 68:243-248

Willoughby EJ (1968) Water economy of the stark's lark and grey-backed finch-lark from the Namib desert of South West Africa. Comp Biochem Physiol 27:723-745

Willoughby EJ (1969) Evaporative water loss of a small xerophilous finch, *Lonchura malabarica*. Comp Biochem Physiol 28:655-664

Willoughby EJ (1970) Composition of avian urine. Science 169:1230-1231

Willoughby EJ (1971) Drinking responses of the red crossbill (*Loxia curvirosta*) to solutions of NaCl, MgC2, and CaC2. Auk 88:828-838

Wilson SC, Lacassagne L (1978) The effects of dexamethasone on plasma luteinizing hormone and oviposition in the hen (*Gallus domesticus*). Gen Comp Endocrinol 35:16-26

Wilson WO (1948) Some effects of increasing environmental temperatures on pullets. Poult Sci 27:813-817

Winget CM, Mepham CA, Averkin EG (1965) Effects of a 17-spirolactone on a circadian rhythm and other physiological systems (*Gallus domesticus*). Comp Biochem Physiol 16:497-506

Wingfield JC, Farner DS (1978) The endocrinology of a natural breeding population of the white-crowned sparrow (*Zonotrichia leucophrys pugetensis*). Physiol Zool 51:188-205

Wolbach RA (1955) Renal regulation of acid-base balance in the chicken. Am J Physiol 181:149-158

Wolfenson D, Berman A, Frei YF, Snapir N (1978) Measurement of blood flow distribution by radioactive microspheres in the laying hen (*Gallus domesticus*). Comp Biochem Physiol 61A:549-554

Wood WG, Ballantyne B (1968) Sodium ion transport and β-glucoronidase activity in the nasal gland of *Anas domesticus*. Anat 103:277-287

Wood-Gush DGM, Horne AR (1970) The effect of egg formation and laying on the food and water intake of brown leghorn hens. Br Poult Sci 11:459-466

Woolley P (1959) The effect of posterior lobe pituitary extracts on blood pressure in several vertebrate classes. J Exp Biol 36:453-458

Wright A, Phillips JG, Huang DP (1966) The effect of adenohypophysectomy on the extrarenal and renal excretion of the salineloaded duck (*Anas platyrhynchos*). J Endocrinol 36:249-256

Wunder BA, Trebella JJ (1976) Effects of nasal tufts and nasal respiration on thermoregulation and evaporative water loss in the common crow. Condor 78:564-567

Yapp WB (1956) Two physiological considerations in bird migration. Wilson Bull 68:312-327

Yarbrough CG (1971) The influence of distribution and ecology on the thermoregulation of small birds. Comp Biochem Physiol 39A:235-266

Yasukawa M (1959) Studies of the movements of the large intestine. VII. Movements of the large intestine of fowls. Jpn J Vet Sci 21:1-8

Young EG, Dreyer NB (1933) On excretion of uric acid and urates by the bird. J Pharmacol Exp Ther 49:162-180

Young EG, Musgrave FF (1932) The formation and decomposition of urate gels. Biochem J 26:941-953

Young EG, Musgrave FF, Graham HC (1933) The influence of electrolytes on the formation and decomposition of urate gels. Can J Res 9:373-385

Yushok W, Bear FE (1948) Poultry manure. Its preservation, deodorization and desinfection. N J Agric Exp Stn Bull 707:3-11

Zeuthen E (1942) The ventilation of the respiratory tract in birds. Biol Medd K Dan Vidensk Selsk 17:1-51

Ziswiler V, Farner DS (1972) Digestion and digestive system. In: Farner DS, King JR (eds) Avian biology, vol II, Chap 6. Academic Press, London New York, pp 343-430

Zucker IH, Gilmore C, Dietz J, Gilmore JP (1977) Effect of volume expansion and veratrine on salt gland secretion in the goose. Am J Physiol 232:R185-R189

Notes Added in Proof

Collins BG, Gayle C, Payne S (1980) Metabolism, thermoregulation and evaporative water loss in two species of Australian nectar-feeding birds (Family *Meliphagidae*). Comp Biochem Physiol 67A:629-635

Curtis MJ, Flack IH (1980) The effect of *Escherichia coli* endotoxins on the concentrations of corticosterone and growth hormone in the plasma of the domestic fowl. Res Vet Sci 28:123-127

Dahm HH, Schramm U, Lange W (1980) Scanning and transmission electron microscopic observations of the cloacal epithelia of the domestic fowl. Cell Tissue Res 211:83-93

Dantzler WH, Braun EJ (1980) Comparative nephron function in reptiles, birds, and mammals. Am J Physiol 239:R197-R213

Duke GE, Bird JE, Daniels KA, Bertoy RW (1981) Food metabolizability and water balance in intact and cecectomized great-horned owls. Comp Biochem Physiol 68A:237-240

Freeman BM, Flack IH (1980) Effects of handling on plasma corticosterone concentrations in the immature domestic fowl. Comp Biochem Physiol 66A:77-81

Freeman BM, Flack IH, Manning ACC (1980) The sensitivity of the domestic fowl to corticotrophin. Comp Biochem Physiol 67A:561-567

Freeman BM, Manning ACC, Flack IH (1980) Short-term stressor effects of food withdrawal on the immature fowl. Comp Biochem Physiol 67A:569-571

Hammel HT, Simon-Oppermann C, Simon E (1980) Properties of body fluids influencing salt gland secretion in Pekin ducks. Am J Physiol 239:R481-R496

Harvey S, Phillips JG (1980) Growth, growth hormone, and corticosterone secretion in freshwater and saline-adapted ducklings (*Anas platyrhynchos*). Gen Comp Endocrin 42:334–344

Harvey S, Merry BJ, Phillips JG (1980) Influence of stress on the secretion of corticosterone in the duck (*Anas platyrhynchos*). J Endor 87:161–171

Karasawa Y (1977) Stimulatory effect of intraportal ammonia on plasma uric acid concentration and urinary uric acid excretion in chickens fed a low protein diet. J Nutr 107:1147–1152

Karasawa Y (1981) Adaptive responses of glutamine, ammonia and uric acid in tissues and urine to ammonia load in chickens fed various levels of dietary protein. Comp Biochem Physiol 68A: 265–267

Kaufman S, Kaesermann H-P, Peters G (1980) The mechanism of drinking induced by parenteral hyperoncotic solutions in the pigeon and in the rat. J Physiol (Lond) 301:91–99

Lingham RB, Stewart DJ, Sen AK (1980) The induction of $(Na^+ + K^+)$-ATPase in the salt gland of the duck. Biochim Biophys Acta 601:229–234

Morley M, Scanes CG, Chadwick A (1980) Water and sodium transport across the jejunum of normal and sodium loaded domestic fowl (*Gallus domesticus*). Comp Biochem Physiol 67A:695–697

Morley M, Scanes CG, Chadwick A (1981) The effect of ovine prolactin on sodium and water transport across the intestine of the fowl (*Gallus domesticus*). Comp Biochem Physiol 68A:61–66

Möhring J, Schoun J, Simon-Oppermann C, Simon E (1980) Radioimmunoassay for arginine-vasotocin (AVT) in serum of Pekin ducks: AVT concentrations after adaptation to fresh water and salt water. Pflügers Arch 387:91–97

Robinzon B, Kare MR, Beauchamp GK (1980) Comparative aspects of salt preference and intake in birds. In: Kare MR, Fregly MJ, Bernard RA (eds) Biological and Behavioral Aspects of Salt Intake. Academic Press, New York, pp 69–81

Sandor T, Mehdi AZ (1980) Corticosteroids and their role in the extrarenal electrolyte secreting organs of nonmammalian vertebrates. In Delrio G, Brachet J (eds) Steroids and Their Mechanism of Action in Nonmammalian Vertebrates. Raven Press, New York, pp 33–49

Schrader C, Weyrauch KD (1976) Lichtmikroskopische, elektronenmikroskopische und histochemische Untersuchungen am Kloakenepithel des Haushuhns. Anat Anz 139:369–385

Simon-Oppermann C, Simon E, Deutsch H, Möhring J, Schoun J (1980) Serum arginine-vasotocin (AVT) and afferent and central control of osmoregulation in conscious Pekin ducks. Pflügers Arch 387:99–106

Wilson JX, Butler DG (1980) The effects of extracellular NaCl, corticosterone and ouabain on the Na^+, K^+, and water concentrations in the nasal salt glands of freshwater and salt-adapted Pekin ducks (*Anas platyrhynchos*). Comp Biochem Physiol 66A:583–591

Systematic and Species Index [1]

Adélie penguin, see *Pygoscelis adeliae*
Agelaius phoeniceus 20
Albatross, see *Diomedea immutubilis* et *nitripes*
Alectoris graeca 37
Amphispiza belli nevadensis 16, 50
Amphispiza bilineata 74
Anthochaera carunculata 98

Barbary dove, see *Streptopelia rivoria*
Barn swallow, see *Hirundo rustica*
Black-throated sparrow, see *Amphispiza bilineata*
Blackpoll warbler, see *Dendroica striata*
Bobwhite, see *Colinus virginianus*
Bonasa umbellus 37
Bubo virginianus 22
Budgerigar, see *Melopsittacus undulatus*

Cacatua roseicapilla 23, 94, 96, 101, 110, 120, 148
Cactus wren, see *Campylorhynchus brunneicapillum*
California quail, see *Lophortyx californicus*
Campylorhynchus brunneicapillum 48
Canachites canadensis 37
Capercailzie, see *Tetrao urogallus*
Carnivorous birds 13, 22, 32, 93, 134, 140, 156
Carpodacus mexicanus 16, 56, 75
Cathartes aura 87, 90
Charadiiformes 137
Chukar partridge, see *Alectoris graeca*
Colinus virginianus 22, 37
Common crow, see *Corvus brachyrrhynchos*
Coot, see *Fulica americana*
Cormorant, see *Phalocrocorax auritus, pelagicus* et *varius*
Corvus brachyrrhynchos 48
Coturnix coturnix 14, 26, 36, 37, 77
Crested pigeon, see *Ocyphaps lophotes*

Dacelo gigas 96, 98
Dendroica palmarum 166
Dendroica striata 167
Double-crested cormorant, see *Phalacrocorax auritus*
Dromaius novaehollandiae 5, 71, 83, 93, 96, 110, 114, 148

Emu, see *Dromaius novaehollandiae*
Eremopterix verticalis 51, 164
Eudyptula minor 83
European starling, see *Sturnus vulgaris*

Falco sparverius 22
Fulica americana 129

Galah, see *Cacatua roseicapilla*
Galapagos mocking bird, see *Nesominus macdonaldi* et *parvulus*
Gambel's quail, see *Lophortyx gambelii*
Geococcyx californicus 5, 35, 57, 97, 169
Glaucous-winged gull, see *Larus glaucescens*
Goose, domestic 35, 59, 129
Grass parrot, see *Neophema bourkii*
Great black-backed gull, see *Larus marinus*
Great horned owl, see *Bubo virginianus*
Grey-backed finch lark, see *Erenopterix verticalis*
Grouse, see *Lagopus* sp
Guam rail, see *Rallus owstoni*

Herring gull, see *Larus argentatus*
Hirundo rustica 3
House finch, see *Carpodacus mexicanus*
Humming birds, see *Trochlidae* sp.

Inca dove, see *Scardafella inca*

Japanese quail, see *Coturnix coturnix*

Kittiwake, see *Rissa tridactyla*
Kookaburra, see *Dacelo gigas*

Lagopus lagopus 34, 39
Lagopus mutus 36, 39
Lagopus sp. 37, 39
Large-billed savannah sparrow, see *Passerculus sandwichensis beldingi*
Larus argentatus 17, 26, 126, 153
Larus fuscus 153
Larus atricilla 17
Larus glaucescens 5, 99, 129, 136, 143
Larus marinus 138, 140
Laughing gull, see *Larus atricilla*
Leach's petrel, see *Oceanodroma leucorrhoa*

1 All birds mentioned in text, except domestic fowl and duck

Lesser black-backed gull, see *Larus fuscus*
Lonchura malabarica   51
Lonchura sp.   77
Lophortyx californicus   16, 20, 75
Lophortyx gambelii   16, 20, 23, 53, 71, 80, 83

Melospiza melodia juddi   74
Meliphaga virescens   94, 98
Melopsittacus undulatus   14, 20, 47, 69, 73, 78, 88, 92, 98, 110, 148, 162
Mountain white-crowned sparrow, see *Zonotrichia leucophrys oricantha*
Mourning dove, see *Zenaidura macroura marginella*

Nandu, see *Rhea americana*
Neophema bourkii   23
Nesomimus macdonaldi et *parvulus*   79

Oceanodroma leucorrhoa   136
Ocyphaps lophotes   98
Ostrich, see *Struthio camelus*

Palm warbler, see *Dendroica palmarum*
Passerculus sandwichensis beldingi, rostratus et *brooksi*   20, 56, 71, 73
Pelagic cormorant, see *Phalacrocorax pelagicus*
Penguin, see *Eudyptula minor*
Phalacrocorax auritus   47, 62, 75, 125, 129, 136
Phalacrocorax pelagicus   47
Phalacrocorax varius   83
Phasianus colchicus   26, 34
Pheasant, see *Phasianus colchicus*
Pigeon, domestic   14, 47, 78, 85, 89, 146, 167, 169
Poephila guttata · 5, 15, 20, 23, 46, 69, 74, 77, 92, 94, 98, 157
Pterocles lichtensteinii, namagua et *senegallus*
Purple cracle, see *Quiscalus quiscala*
Pygoscelis adeliae   9

Quiscalus quiscala   17

Rallus owstoni   129
Red-eyed vireo, see *Vireo olivaceus*

Red wattle bird, see *Anthochaera carunculata*
Red-winged blackbird, see *Agelaius phoeniceus*
Rhea americana   34, 57, 93
Rissa tridactyla   143
Roadrunner, see *Geococcyx californicus*
Rock ptarmigan, see *Lagopus mutus*
Ruffed grouse, see *Bonasa umbellus*

Sage sparrow, see *Amphispiza belli nevadensis*
Savannah sparrow, see *Passerculus sandwichensis beldingi, rostratus* et *brooksi*
Scardafella inca   52
Senegal dove, see *Streptopelia senegalensis*
Senegal sandgrouse, see *Pterocles senegallus*
Silver-billed finch, see *Lonchura malabarica*
Singing honeyeater, see *Meliphaga virescens*
Song sparrow, see *Melospiza melodia juddi*
Sparrow hawk, see *Falco sparverius*
Spizocoryx starki   51, 164
Spruce grouse, see *Canachites canadensis*
Stark's lark, see *Spizocoryx starki*
Streptopelia rivoria   46, 52
Streptopelia senegalensis   96, 98
Struthio camelus   41, 46, 52, 57, 71, 93, 136, 164
Sturnus vulgaris   17, 53, 71, 90, 168

Tetrao urogallus   34, 93
Trochlidae sp.   41
Turkey, domestic   26, 30, 36, 73, 76, 91, 104, 106
Turkey vulture, see *Cathartes aura*

Vireo olivaceus   166

White-crowned sparrow, see *Zonotrichia leucophrys gambelii* et *pugetensis*
White-throated sparrow, see *Zonotrichia albicollis*
Willow ptarmigan, see *Lagopus lagopus*

Zebra finch, see *Poephila guttata*
Zenaidura macroura marginella   16, 20
Zonotrichia albicollis   79
Zonotrichia leucophrys gambelii   14, 77, 166
Zonotrichia leucophrys oricantha   169
Zonotrichia leucophrys pugetensis   151, 166

Subject Index

Acid excretion in kidney   72
Active transport in coprodeum-colon   107, 116
Adrenal cortex
  weight after salt loading   128
Adrenalectomy   151, 154
Aldosterone   122, 151
Amiloride   118–120, 123
Amino acids
  effect on cloacal resorption   120, 123, 150
  in urine   72
Ammonia
  cloacal absorption   111, 114, 149
  renal excretion   72, 86
Angiotensin   124, 154
Angiotensin-induced drinking   14
Antidiuresis   76, 80
Arginine vasopressin   75
Arginine vasotocin   75
  egg-laying   170
  renal effects   80

Blood pressure
  effect of AVT   83
Body weight
  determination of   3

Caecum   32, 93, 97
Chloride concentration
  in budgerigar plasma   162
  in contents of the small intestine   33
  in plasma   8
  in salt gland fluid   132
  in urine   70
Chloride transport in coprodeum-colon
  in vitro   118
  in vivo   109
Cloaca
  absorption, direct determination   107
  absorption, indirect estimates   101
Collecting duct   56
Colon epithelium   95
Coprodeum
  storage of urine   97
Corticosterone   140, 151, 170
Corticotrophin (ACTH)   153
Cortisol   154

Counter-current system
  exchange in airways   47, 156
  in kidney   73
  in salt gland   137

Dehydration   21, 72, 77
  calculation of cloacal resorption   145
  cloacal absorption   112
  concentration of AVT   83
  effect on exteriorised birds   102
  effect on urine regurgitation   97
  salt gland secretion   129
Desert adaptation   47, 196
Diabetes insipidus   78
Digestion in caecum   38
Divalent ions
  in salt gland fluid   136
  in urine   72, 88
Drinking rate   18
  during egg-laying   171
  in zebra finch   160

Egg-laying   83, 170
Electrical potential difference   31, 107, 108, 112, 116, 120
Eminentia medialis   77
Energy
  costs of migration   167
Evaporation
  during flight   163
  from skin   46
  in exteriorised birds   102
  in relation to output from cloaca and salt gland   142
  reaction to dehydration   51, 159, 162, 164
  relation to ambient temperature   50
  relation to body weight   43
  relation to oxygen uptake   41
Exchangeable sodium   5
Extracellular fluid volume   5

Faeces
  swallowing of faecal sacs   169
  water content   38, 99, 164

Glomerular filtration rate (GFR)   59, 147, 154
   effect of AVT   80
   exteriorised birds   103
   salt loading   62
Glucose   37, 117
Gular flutter   51, 46
Gut water   31

Heat loss
   influence of dehydration   51
   evaporative   163
Hypophysectomy   154
Hypothermia
   nocturnal   52

Inulin   9, 81, 85, 91, 105

Juxta-glomerular apparatus   154

K, see potassium
Kidney structure   74
K_m values   107, 112, 119, 123, 150

Large intestine
   storage of urine   97
Lysine vasopressin   76

Metabolic water production   41, 161, 167
Migration   165

Na, see also sodium
Na-K-ATPase   118, 127, 130, 138
Na-gradient hypothesis   120
Na-influx across luminal membrane   118
Nasal salt glands   125
Nephron   64, 75
Neurohypophysectomy   78, 170
Nitrogen excretion   84, 125, 149

Osmoreceptors   129
Osmotic permeability coefficient   107, 112, 163
Oviduct   76, 103
Oviposition   170
Oxygen consumption   40, 161, 163
Oxytocin   75, 170

p-aminohippurate   154
Panting   51
Permeability to salt   120
Phosphate
   concentration in urine   70
   absorption in cloaca   114, 149
Pitressin   76
Plasma osmolality   9
Plasma sodium concentration   9
Plasma volume   9
Polydipsia   79, 101, 102, 104, 173

Polyethylene glycol   30, 35, 81
Portal blood flow in kidney   54–57
Potassium concentration
   in contents of the small intestine   33
   in salt gland fluid   132
   in urine   70
Potassium secretion
   cloacal   114, 149
   renal   71
Precipitation of uric acid   87
Premigratory fattening   166
Prolactin   153

Reflection coefficient   30, 114
Renal plasma flow   58, 154
Renal water excretion   63
Renin-angiotensin system   154
Respiratory quotient (RQ)   163
Respiratory water loss   47
Retrograde flow of urine   34, 93, 97, 145

Salinity tolerance   23
Salt depletion   28, 102, 112
Salt glands   125
   effect of corticosteroids   140
   secretion stimulus   129
Salt loading   28
   effect on cat-ion trapping   90, 112, 113
   effect on salt glands   139
   secretion of corticosterone   152
Salt poisoning, in poultry   27
Salt-water drinking   24, 161
Short-circuit current of isolated coprodeum
   116, 210
Single-nephron glomerular filtration rates   64
Sodium
   appetite   14
   concentration in plasma   4
   depletion   28
   in contents of the small intestine   33
   in urine   67
   in salt gland fluid   132
   inflow across luminal membrane   118
   permeability   118
   reabsorption in kidney   63
Solute-linked water flow   30, 107, 109, 145, 163
Spironolactone   31
Supraoptic nucleus   79

Taste for saline   16
Tears, salt excretion in   142
Thirst   14
Total body water   3, 166, 171
Transepithelial potential difference, see electrical
   P. D.
Tritiated water   3, 106, 108, 171
Tubules, kidney secretion and excretion   63

Unstirred layers, transport in colon 145
Urea
  clearance of 91
  content in urine 86
Uric acid, excretion 85
Urinary bladder, in rhea americana 93
Urine
  flow rate 63, 80
  nitrogenous constituents 86
  osmolality 23, 69, 80, 147, 160
Ussing chamber 113, 115, 123
Uterus (shell gland) see Oviduct.

Vagus nerve 129
V_{max}-values 108, 112, 119
Volatile fatty acids 32, 37, 39
Volume regulation, salt gland secretion 130

Water absorption
  in caecum 36
  in kidney 63
  in salt gland 138
Water load 65
Water turnover 3
  in carnivorous birds 22
  in zebra finch 161

Zoophysiology

formerly
Zoophysiology and Ecology
Coordination Editor: D. S. Farner
Editors: W. S. Hoar, B. Hoelldobler, H. Langer,
M. Lindauer

Volume 1
P. J. Bentley

Endocrines and Osmoregulation

A Comparative Account of the Regulation of Water
and Salt in Vertebrates
1971. 29 figures. XVI, 300 pages
ISBN 3-540-05273-9

Volume 2
L. Irving

Arctic Life of Birds and Mammals

Including Man
1972. 59 figures. XI, 192 pages
ISBN 3-540-05801-X

Volume 3
A. E. Needham

The Significance of Zoochromes

1974. 54 figures. XX, 429 pages
ISBN 3-540-06331-5

Volume 4/5
A. C. Neville

Biology of the Arthropod Cuticle

1975. 233 figures. XVI, 448 pages
ISBN 3-540-07081-8

Volume 6
K. Schmidt-Koenig

Migration and Homing in Animals

1975. 64 figures, 2 tables. XII, 99 pages
ISBN 3-540-07433-3

Volume 7
E. Curio

The Ethology of Predation

1976. 70 figures, 16 tables. X, 250 pages
ISBN 3-540-07720-0

Volume 8
W. Leuthold

African Ungulates

A Comparative Review of Their Ethology and
Behavioral Ecology
1977. 55 figures, 7 tables. XIII, 307 pages
ISBN 3-540-07951-3

Volume 9
E. B. Edney

Water Balance in Land Arthropods

1977. 109 figures, 36 tables. XII, 282 pages
ISBN 3-540-08084-8

Volume 10
H.-U. Thiele

Carabid Beetles in Their Environments

A Study on Habitat Selection by Adaptations in
Physiology and Behaviour
1977. 152 figures, 58 tables. XVII, 369 pages
ISBN 3-540-08306-5

Volume 11
M. H. A. Keenleyside

Diversity and Adaptation in Fish Behaviour

1979. 67 figures, 15 tables. XIII, 208 pages
ISBN 3-540-09587-X

Springer-Verlag
Berlin
Heidelberg
New York

Two related journals

Zoomorphology

An International Journal of Comparative
and Functional Morphology

ISSN 0340-6725   Title No. 435

Editors: P. Ax, Göttingen, FRG; W. Bock,
New York, NY, USA; R. B. Clark, Newcastle
upon Tyne, Great Britain; R. Eakin, Berkeley,
CA, USA; O. Kraus (Managing Editor),
Hamburg, FRG; G. Kümmel, Karlsruhe,
FRG; R. M. Rieger, Chapel Hill, NC, USA;
V. Ziswiler, Zürich, Switzerland

Zoomorphology, an international journal of
comparative and functional morphology,
accepts original papers based on research in
animal morphology at all levels of ontogeny
and organization including ultrastructure.
Emphasis will be placed on the following
fields:

- The studies of homologies and develop-
  mental phenomena as the basis for phylogenetic relationships and systematic.

- Interrelationships between structure,
  general biology, and environment.

- Functional morphology: synthesis of
  structure and function.

Subscription Information and/or sample
copies are available upon request. Please send
your order or request to your bookseller or
directly to: Springer-Verlag, Journal Promo-
tion Department, P. O. Box 105280,
D-6900 Heidelberg, FRG

Springer-Verlag
Berlin
Heidelberg
New York

Journal of Comparative Physiology · A+B

Founded in 1924 as
Zeitschrift für vergleichende Physiologie
by K. von Frisch and A. Kühn

A. Sensory, Neural, and Behavioral
Physiology
ISSN 0340-7594   Title No. 437
Editorial Board: H. Autrum, R. R. Capranica,
K. von Frisch, G. A. Horridge, M. Lindauer,
C. L. Prosser
In cooperation with a distinguished board of
advisory editors.

B. Biochemical, Systemic, and Environmental
Physiology
ISSN 0174-1578   Title No. 437
Editorial Board: K. Johansen, B. Linzen,
W. T. W. Potts, C. L. Prosser
In cooperation with a distinguished board of
advisory editors.

The Journal of Comparative Physiology pub-
lishes original articles in the field of animal
physiology. In view of the increasing number
of papers and the high degree of scientific
specialization the journal is published in two
sections.

**A. Sensory, Neural, and Behavioral Physio-
logy**
Physiological Basis of Behavior; Sensory Phy-
siology; Neural Physiology; Orientation,
Communication; Locomotion; Hormonal
Control of Behavior

**B. Biochemical, Systemic and Environmental
Physiology**
Comparative Aspects of Metabolism and
Enzymology; Metabolic Regulation; Respira-
tion and Gas Transport; Physiology of Body
Fluids; Circulation; Temperature Relations;
Muscular Physiology